ENVIRONMENTAL TOXICOLOGY

ENVIRONMENTAL TOXICOLOGY

By

M. Satake
Y. Mido
H. Yasuhisa
S. Taguchi
M.S. Sethi
S.A. Iqbal

DISCOVERY PUBLISHING HOUSE PVT. LTD.
NEW DELHI-110 002

First Published - 2001

Reprinted - 2016

ISBN: 978-81-7141-350-8

Environmental Toxicology

Published by:

DISCOVERY PUBLISHING HOUSE PVT. LTD.

4383/4B, Ansari Road, Darya Ganj

New Delhi-110 002 (India)

Phone: +91-11-23279245, 23253475, 43596065

E-mail: discoverypublishinghouse@gmail.com

sales@discoverypublishinggroup.com

web: www.discoverypublishinggroup.com

Printed at:

Infinity Imaging Systems

Delhi

Preface

The teaching of Chemistry at the introductory stage becomes each day a more challenging task as the subject matter becomes more diverse and more complex. These challenges have evoked a series of responses—the present set of introductory chemistry monographs is one such. The teaching of chemistry recognises a number of problems that confront those who select text books. In order to overcome these problems, this volume **"Environmental Toxicology"**—one of about fifty in the Chemistry Monograph Series—is introduced. Each volume is independent of the others deals with one of Chemistry topics and constitutes and complete entity. Each volume is more comprehensive than can be possible in a single volume text. It is intended to provide a range of topics to cover most undergraduate and chemistry main courses of study. These volumes can be used to enrich the more conventional courses of study.

Suggestions for improvement are welcome and shall be gratefully acknowledged.

Authors

Contents

Contents

1

Introduction to Toxicology

Linnaeus first introduced the classification of living organisms into species and laid the foundation for the system used today in which each organism is given a generic and specific name. The environment has been defined as all the inanimate compartments such as air, soil and water and all the animate components such as plants and animals that surround a particular organism or population of organisms.

Toxicology is the study of the adverse effects of the chemicals and physical agents on living organisms. In classical toxicology the test animal is the laboratory rat.

Not only can the toxicant directly affect the organism being evaluated, but also the toxicant can adversely affect both biotic and abiotic parameters such as the prey or source of food of the species being studied, thus having an indirect effect on the survival of the organism. As a result of the complex simultaneous interaction between the toxicant in question on many species and the variation in the outcome of these effects on different species, many toxicological studies have resorted to studying the effect of a toxicant on only one species at a time. Thus, studies to evaluate the effect of pollutants on man are often included.

In evaluating the toxicity of a compound on man using classical toxicological approaches, surrogate species such as rodents are chosen. In many respects surrogate species are often physiologically different from the species on which data is actually required. Therefore, there is always some question about their validity or the applicability of extrapolating from one species to another. Even though a mammalian toxicity test may be designed to determine the effects of a compound on more than one species, extrapolations are necessary for only a small number of species.

For many years the prime criterion for selecting a specific organism for toxicity testing was its high degree of sensitivity to chemical compounds. Emphasis is shifting to the use of those organisms that are the most reliable predictors of the responses of the organisms or even of communities or ecosystems. Also, many species that are extremely sensitive or have high correspondence with the responses of other organisms are exceedingly difficult to maintain the laboratory. Thus, it is more reasonable to use test organisms that have been widely for toxicity testing and whose strengths and weaknesses for this purpose are well known.

A major thrust as present is the evaluation of chemicals found at hazardous waste sites and on human and ecological health.

Modern insecticides are extremely potent, and not only do they poison the target species but often persist in the environment and are accumulated in nontarget organisms, including fish and wildlife. The bioaccumulation potential of lipophilic insecticides tends to be inversely related to their rates of metabolism. Metals are also bioaccumulated by plants and animals. For example, certain aquatic plants bioconcentrate selenium 500 times from the surrounding water. Subsequent consumption of these plants by invertebrates and forage fish can result in additional biomagnification of 2 to 6 times the plant concentration.

About 63,000 chemicals are in common use worldwide, with 3000 compounds accounting for almost 90% of the total production. About 200 to 1000 new synthetic chemicals are marketed each year. Few ecological assessment on the bulk of the materials on the market or those introduced each year have been undertaken. This need was perceived as very important, thus many countries have promulgated legislation that requires both toxicological and ecotoxicological testing of chemicals.

Analytical instruments cannot measure toxicity. They can measure only the concentration of a chemical in the environment or in organism. The importance of using living organisms to assess toxicity, rather than attempting to assess the toxicity of a compound from its chemical concentration in the environment or in an animals, is being felt now. However, both chemical and physical measurements are still extremely important in assessing toxicity, because the concentration of the toxicant available to the test organisms must be known.

The analyst who is attempting to determine the concentration of a pollutant in an environmental sample has to face several problems. Difficult arise when we have to distinguish between chemical species, ions, molecules, or complexes and the fraction in which the species is found such as the soluble fraction. With new techniques some of the difficulties in the interpretation of the concentration of metals in the environment may be overcome. The chemical form and species of a metal can to a large extent affect the metal's behavior and its toxicological effects. For example, the different forms of selenium have different toxicological potencies, with selenite being more toxic to rats in drinking water than selenate. The presence of natural components such as humic acids and particulates in suspension to which metals can adsorb.

Chemicals not toxic to humans and animals can alter abiotic factors resulting in enormous repercussions either at the local level or worldwide. These effects could also be vital to man's survival. For example, the increase in the concentration of both carbon dioxide and chlorofluorocarbons (CFCs) in the atmosphere may have far reaching consequences. Carbon dioxide had a mean natural concentration at the beginning of the century of about 0.028%. Since then its emission through the combustion of fossil fuels have been increasing by about 0.002% each year. This continual increase in the carbon dioxide is predicted to elevate global temperatures, which could :

1) affect the distribution of certain crops.

2) enhance the melting of the ice caps which could raise the sea level, resulting in submersion of many low lying cities of the world.

3) also enhance the survival of certain pests, resulting in decreased crop production and many modify the adaptation of C3 and C4 plants.

The net results will be radical alterations in the distribution of plant species in the world. CFCs, developed due to their stability and low toxicity, have contributed to the depletion of stratospheric ozone, which is increasing the risk of skin cancer and damage to vegetation.

Other pollutants such as DDT and acid rain, which are both relatively harmless or only slightly toxic to mammals and man, have direct and indirect effects on an ecosystem. DDT jeopardized the survival of eagles and other birds at high trophic levels by interfering with the normal function of estrogen, the hormone involved in the deposition of calcium in the egg shell. Acid deposition is thought to be the major cause of the destruction of populations of fish and other aquatic organisms in poorly buffered lakes. Fish die from failure of their self-regulating mechanisms, although tests have shown that the pH of the water has to drop below 4.6 before a physiological effect is noted. Nevertheless, some sensitive fish species have disappeared from lakes with pH readings of 5.0. The increased toxicity may be due in a large part to the leaching of aluminum from certain solids by acid rain, although other physical and chemical factors may also be important.

In order to determine the impact of important in the environment, we must be aware of other changes that are occurring around us. Despite the large amount of data available on some pollutants, quantitative information on sources is often not available.

This text has been complied to deal with many of the above mentioned effects of pollutants on man, animals and the environment, and has been divided into four section. The first chapter given an introduction of toxicology and Chapter 2 through present the effects of difficult type of toxicants.

2

Basic Principles of Toxicology

THE POLLUTANTS

A pollutant or toxicant is defined as a substance that occurs in the environment, at least in part as a result of an anthropogenic discharge from industrial activity, and had a deleterious effect on living organisms. A *contaminant* on the other hand is a compound released by man's activities that does not necessarily have a deleterious biological effect. Frequently, the term pollutant is used loosely to cover both definitions. A pollutant may kill more than 50% of the individuals of a species in a population but be of little or no ecological important unless this loss has a direct effect on the survival of the ecosystem. Some pollutants, in the amounts discharged into the environment, may not have a direct effect on living organisms, but they may alter the physical and chemical environment, impairing the survival of a species. For example, runoff of inorganic nutrients, particularly phosphorus and nitrogen derived from sewage and agricultural fertilizers into water bodies, enhance the multiplication of algae, a process known as *eutrophication*, resulting in a massive requirement for oxygen by these rapidly growing plants. As a result, the oxygen concentration in the water body is decreased and many fish species may not survive.

In order to understand the effect of pollutants on organisms, we need to understand how the compound is absorbed, how it is distributed by the blood stream (if the organism has one), its subsequent metabolism, and its excretion. During these four basic phases of the passage of a pollutant through an organism (knows as *pharmacodynamics*) several metabolic processes in the animal or plant can be altered. The outcome of these processes can be exemplified by either a drastic response such as death or a less total effects but one in which the physiological responses of the animal or plant are dangerously altered. Death usually occurs when the organism is exposed for a short time to a high concentration of a compound. Exposure to lower concentrations over a longer period may result in a subtle biochemical or physiological effect which in itself does not appear to affect the health of the organism. However, extended periods of exposure could result in changes that can initiate the beginning of a disease or condition that could affect the survival of an individual or a population.

ROUTES OF EXPOSURE TO TOXINS

Before a substance can injure on organism, it must contract or enter the organism (radioactive materials and explosives are exception). Foreign substances enter by four possible routes;

1. inhalation,
2. skill (or eye) absorption
3. ingestion, and
4. injection.

Often the mode of entry is significant for the degree of the harm that a foreign agent causes. For example, although rubbing alcohol (isopropyl alcohol) can be beneficial when applied to the skin; ingesting it causes serious illness. Thus, decisions regarding exposure must involve careful attention to the probable route of entry.

Inhalation

For human, inhalation is the most frequent route of exposure to chemicals. A person inhales about 20,000 liters of air per day. Thus, even moderate levels of air pollution can result in significant exposure. Furthermore, lung tissue is highly susceptible to the action of chemical agents such as acidic and caustic vapours, which can destroy cells. Even chemical agents that do not directly affect the lungs may pass through them into the bloodstream. For example, inhaled carbon monoxide (CO) does not damage lung tissue directly; rather, it enters the bloodstream, where it attaches to hemoglobin molecules and severely impairs the blood's oxygen-carrying capacity. The brain is particularly sensitive to reduced levels of blood oxygen. As a consequence, inhaling concentrations can cause loss of consciousness and eventually death.

Another odourless gas is methane (CH_4), the primary component of natural gas. Although methane is not directly poisonous, it is explosive. To warn natural gas users of a possible leak, trace amounts of a quite odorous gas (ethyl mercaptan), are intentionally added to natural gas.

The water solubility of gas or vapour is an important factor in determining the amount of inhaled material that reaches the lung. Water-soluble gases dissolve readily in the moisture associated with the mucous membrane of the nose and upper respiratory tract. This causes irritation at these sites. At low airborne concentration, relatively little of these substances will reach the lungs owning to the "scrubbing" effect.

At high atmospheric concentration, however, some of the gas or vapour will not be absorbed at these upper respiratory sites, and amounts sufficient to cause server irritation and pulmonary edema can reach the alveoli.

Comparatively insoluble gases, such as nitrogen dioxide are not removed by the moisture in the nose and upper respiratory tract and can easily reach the terminal recesses of the lung. Substances of intermediate solubility, such as ozone cause irritation of both the upper respiratory tract and the lung.

Gases such as carbon monoxide do not irritate the respiratory tract but are rapidly absorbed into the blood, resulting in systemic intoxication.

The particle size of aerosols determined the extent of their accessibility to small airways. If the diameter of the particles is larger than 10µ, impaction occurs on the mucous coat of the pharynx or nasal cavity, and the particles do not reach the alveoli. For this reason, particles of 10μ or less in diameter are termed respirable. Particles between 1μ and 5μ often sediment within the bronchioles, whereas particles less than 1μ can diffuse within the alveoli.

Particulate matter may deposit on the ciliary epithelium, which covers the upper respiratory tract down to the level of the terminal bronchioles. These particles, either in a free state or after phagocytosis, are moved by cilia and the mucous blanket toward the glottis, where they are swallowed or expectorated. Particulate matter that deposits beyond the ciliary epithelium may be absorbed through the alveolar lining into the blood, or as free and phagocytized particles,, may enter the lymphatic system. Several industrially important substances, such as crystalline silica, beryllium, and asbestos, resist solubilization by the blood or removal by phagocytes, and remain in the alveoli indefinitely, Irritation, inflammation, fibrosis, allergic sensitization, or malignant change may occur. However, substances such as iron oxide can be present for extended period of time with no apparent ill effects.

Absorption

A second route of exposure is direct contact with skin and eyes. Some chemicals such as strong acids (e.g. sulfuric acid [H_2SO_4] in automobile batteries) or alkaline materials (e.g., sodium hydroxide [NaOH] in drain cleaners) destroy (Burn) tissue directly and are particularly painful. Other chemicals pass through the skin into the bloodstream and are transported to other organism where damage may occur. Some pesticides (organophosphates) enter the body in this way and impair the ability of nerves to transmit messages. Death can result if enough of a chemical is absorbed. Skin abrasions and cuts allow a more direct entry for chemicals. The eye, unlike dry skin, is particularly vulnerable because many air pollutants readily dissolve in the fluid on the eye's surface.

When a substance contacts the skin, four actions are possible.

1. The skin and its associated film of lipid and sweat may act as an effective barrier that the substance cannot penetrate.
2. The substance may react with the skin surface and caùse primary irritation.
3. The substance may penetrate the skin and cause sensitization.

4. The agent may penetrate the skin, enter the blood, and act systemically.

Only a small number of substances are absorbed through the skin in hazardous amounts. In order to pass in to the skin, the substance must enter through one or more of the following four routes: the epidermal cells, the sweat glands, the sebaceous glands, or the hair follicles. The pathway through the epidermal cells and the overlaying stratum corneum into the blood is probably the main avenue of penetration, since this tissue constitutes the majority of the surface area of the skin. The stratum corneum plays and critical role in determining cutaneous permeability. Absorption is faster through skin that is abraded or inflamed. For this reason chemicals that are not normally considered hazardous may be dangerous to persons suffering from active inflammatory dermatoses.

Ingestion

A third route of exposure is ingestion through consumption of contaminated food or beverages or by accidental means. By entering food webs, environmental containments such as pesticides and heavy metals can eventually get into the food we eat. **Heavy metals** include the elements lead (Pb), mercury (Hg), chromium (Cr), nickel (Ni), cadmium (Cd), and arsenic (As). Humans and other species that occupy upper trophic levels can ingest significant quantities of toxins as a result of **bioaccumulation**, the process whereby certain chemicals become more highly concentrated as they are transferred become more highly concentrated as they are transferred to organisms at successively higher tropic levels.

Most foods also contain low levels of natural toxins. Chocolate contains theobromine, a possible gene-altering substance; potatoes contains solanine and chaconine, which can reach lethal levels if the potatoes are damaged or exposed to light for extended periods; celery, parsley, parsnips and figs contain psoralen, derivatives, which become active cancer-causing agents when activated by light. Mushrooms, black pepper, mustard, horseradish, cottonseed oil, many herbal tests, and peanut butter all contain small amounts of at least one component with toxic properties. When we ingest these substances we knowingly or unknowingly encounter the small risk associated with exposure to them.

Most ingested toxins must be absorbed by the digestive tract before they initiate any toxic response, but a few chemicals burn or irritate the gastrointestinal tract directly. For example, accidental ingestion of strong alkalis or strong acids severely burns the mouth and even the stomach. Swallowing toxic substances is the most common cause of poisoning among young children. On the other hand, many therapeutic drugs are administered orally because they are absorbed through the gastrointestinal tract and then distributed throughout the body.

Our homes and offices contain a variety of potentially hazardous chemicals. Of particulars concern is preventing the accidental exposure of small children.

Injection

The fourth route of exposure to chemicals is injection through the skin by needle and syringe. Injection is often the preferred means of administering medications because it is most direct route to target organs. While accidental injection is an unlikely route of exposure for the general public, medical personnel are at considerably greater risk to such exposure. Accidental puncture wounds from used needles contaminated with the Acquired Immune Deficiency syndrome (AIDS) virus or potent drugs (for example, drugs used to treat cancer) can expose doctors, nurses, paramedics, and sanitation staff to life-threatening agents.

Lethal and Sublethal Effects

Following exposure to a pollutant, some species in a community may be able to survive and maintain their numbers, but in other cases populations in a community may be adversely affected resulting in the death of that species. Thus some species can adapt to adverse condition while other cannot. With continuous exposure to the pollutant in question adaptation and tolerance may be affected, so that sublethal responses are manifest which could affect the long term survival of the species. Eventually the concentration of the pollutant may reach a high enough value to cause the death of the members of a population. Thus the effects of a pollutant on an organism can be divided into two phases, a lower concentration that result in a sublethal effect which may be detected or a lethal response in which the organism dies. Some sublethal effects may be visually undetectable but with physiological or biochemical means can be identified.

RESPONSES TO TOXIN EXPOSURE

The body's response to foreign chemical—whether inhaled, absorbed, ingested, or injected— depends on a host of factors. They include the frequency and duration of exposure, the dose, and the age, and sex and general health of an exposed individual. Some effects of pollutant on humans and others mammals are given in table 2.1.

The earliest changes in an organism following exposure to a pollutant occurs at the cellular level. Some of the more important effects are alteration in the structural components of the cell membrane, inhibition of certain enzyme such as microsomal enzymes, interferences in protein, lipid, or carbohydrate biosynthesis or metabolism, and alteration in DNA fidelity resulting in mutations and interference with the regulation of cell required for cellular precesses including catabolism, anabolism, and reproduction. The mechanism of the initial changes at the molecular level is known in some cases. Alterations in the normal functions of the cell membrane can occur as a result of alterations in the synthesis of phospholipid or glycoprotein components. Mercury alters the conformation of proteins by interaction with sulphydryl (-SH) groups of cysteine residues resulting in the formation of sulphydryl bridges between two SH groups. Toxic compounds can also interfere with the synthesis of specific macromolecules. For example, lead interferes with the synthesis of hemoglobin by inhibiting two enzymes in the heme biosynthesis pathway. Toxic compounds may bind covalently with DNA, thus rendering it inactive or impairing its normal function.

Certain toxicants are also known to interfere with metabolic pathways crucial to the survival of the plant or animal. These effects may occur as a result of competitive or noncompetitive inhibition. Fluoroacetate, a compound found in a South African plant, has been used as rodent bait. This compound replaces acetate in the condensation reaction between acetate and oxaloacetate, and forms fluorocitrate. This compound blocks aconitase in the Krebs, cycle, thus inhibiting the catabolism of acetic acid derived from glucose and the subsequent generation of energy.

Table 2.1 : Effects of Pollutants on Human and Other Mammals

Biochemical effects

1. impairment of enzyme function by the binding of the toxicant to enzymes, coenzymes, metal activators, or enzyme substrates.
2. alteration of cell membrane or carries in cell membranes.
3. interference with lipid metabolism, resulting in excess lipid accumulation
4. interference with respiration
5. interference with carbohydrate metabolism
6. stopping or interfering with protein biosynthesis through toxic effects on DNA.
7. interference with regulatory processes mediated by hormones or enzymes

Clinical response to toxicants

1. alterations in the vital signs of temperature, pulse rate, respiratory rate and blood pressure.
2. abnormal skin colour
3. effects on the eye, which include:
   — miosis (extensive contraction of pupil)
   — mydriasis (excessive pupil dilation)
   — conjuctivities (inflammation of the membrane covering the front of the eyeball)
   — nystagmus (involuntary movement of the eyeballs)
4. gastrointestinal effects: pain, vomiting, paralytic ileus (stoppage of normal peristalsis)
5. central nervous system effects: convulsions, paralysis, hallucinations ataxia and coma

Sub-clinical effects of toxicants

1. damage to the immune system
2. chromosomal abnormalities
3. modification of the functions of liver enzymes
4. slowing of the conduction of nervous impulses

In most ecosystems animals and plants are exposed to many pollutants simultaneously rather then to a single pollutant. Interaction between these pollutants can enhance or decrease the toxicity of the mixture. The mode of interaction between different pollutants can be due either to the chemical structure of the molecules or to alterations in the physiological processes with the organism. Thus the uptake, distribution, metabolism, storage, or excretion of the compound could be altered. Exposure to two or more compounds can result in an additive, inhibitory (decreased), or *synergistic* (increased) response. An additive response is often observed when two organophosphate compounds are given simultaneously. An increased or synergistic response occurs when the combined effect of two compounds is greater than the sum of the individual compounds. Such a response has been observed in mammals when the hepatotoxins, ethanol, and carbon tetrachloride are administered together to rats. *Potentiation* occurs when one of the two compounds to which the organism is exposed is non-toxic alone but when given with the second compound enhances its toxic response. Isopropanol alone is non-toxic, but when administered with carbon tetrachloride it enhances the latter's toxicity. *Antagonism* occurs when two compounds are administered together or sequentially and one interference with the action of the second. An antagonistic interaction between two pollutants can occur when the two compounds interact at the same site in the target tissue.

Interaction between a pollutant and a natural chemical in the environment can result in the formation of new molecules or complexes, thus preventing the expected uptake of a compound by an organism. For example, in an aquatic environment metals are absorbed by organisms oxides coprecipitate cobalt, zinc, and copper. In the presence of these oxides, the absorption of soluble cobalt, zinc, or copper could thus be inhibited.

Duration and Frequency of Exposure

Chemical exposures are generally classified as either acute or chronic. **Acute exposure** is short-term exposure to a relatively high concentration of a toxin. It usually occurs as a consequence of transportation or industrial accidents. Symptoms of acute exposure usually appear immediately and are often life-threatening. A particularly tragic incident occurred in Bhopal, India, in December 1984 when one of the gaseous chemical ingredients MIC used to manufacture is pesticide was released accidently. A toxic cloud swept downwind from the chemical plant into a densely populated neighborhood and within a few hours killed more than two thousand people and injured over ten thousand others.

Chronic exposure involves repeated or continuous exposure to relatively low levels of toxins For example, cigarette smokers and people who live or work with smokes daily inhale low amounts of tars and nicotine released by burning tobacco. Damaged from chronic exposure often appears only after many years. Serious problems develop more frequently in certain industrial occupation. Long-terms exposure to high levels of dust in occupation such as mining, metal grinding, and the manufacture of abrasives, for example, may eventually cause lung disease such as black lung (from coal dust); byssinosis, or brown lung (from cotton dust); silicosis (from quartz dust generated during mining); and asbestosis (from asbestos fibers).

Dose-Response Relationships

We all know that a single aspirin tablet does not usually produce a harmful effect, but swallowing the entire contents of a small bottle of aspirin at once may be lethal. The size of the dose is an important determinant of the specific response. **Toxicity** is the type and intensity of response evoked by a chemical. To determine the response to chemicals, toxicologists administer controlled doses to laboratory test animals, usually rats or mice, and use the information they gain from these experiments to approximate the hazards for humans.

The toxic effects of chemicals are manifested in various ways Table. 2.1. Some chemicals interfere with the ability of an organ (e.g., the liver or kidney) to function. Toxins may also interfere with blood-forming mechanisms, or the functioning of **enzymes** (proteins that accelerate specific chemical reactions), the central nervous system (including the brain), or the immune system. For example, dioxin, an extremely toxic compound, is thought to affect DNA and ultimately the immune system.

In determining the relationship between dose and response, it is necessary to distinguish between the dose or environmental concentration and the amount of the chemical that reaches the target tissue. In some cases the concentration of a compound may be high in the environmental medium, particularly if it is water soluble, but it the compound is not absorbed and does not reach the target organ, no effect will be observed. It is also necessary to distinguish between response and effect. Dope response is usually the relationship between a measurable physical, chemical , or biological reaction following exposure to a certain concentration of a compound. As it is extremely difficult in most experiments to determine the dose reaching the target tissue, the environmental concentration or the amount of compound administered is used as the dose. The reaction to the dose, or the response that is observed, can be quantitatively measured either by the magnitude of the response or the time taken for a specific response to be recorded.

In measuring the dose response, it is assumed that the response observed is due to administration of the chemical and thus there is a casual relationship. It is also assumed that the response is related to dose and thus there is a particular molecular side with which the chemical reacts, that the response is proportional to the concentration of the compound at the site of action, and that the concentration of the compound is related to the dose administered. Thus, the concentration at the site of action is a function of the dose.

Toxicologists analyze both sublethal and lethal responses to determine toxicity. To measure sub-lethal responses, they monitor blood, urine, organ tissue, and the general health of animals. Sublethal responses may include changes in the number and type of blood cells, alterations in blood chemistry, changes in body weight, tumor formation, and retardation of growth.

Scientists usually find that low doses produce no response (Fig. 2.1). The highest dose that products no toxic response is called the **no-observed-effect level** (NOEL). Doses below this level are considered relatively safe for test animals. As the dose is increased above the NOEL, some test animals begin to show symptoms of poisoning. At some higher dose, a few animals die. At increasingly higher doses, more animals die, until eventually a dose is reached that kills all the animals in the test population.

From dose-response experiments, toxicologists determine a common measure of the relative toxicity of different toxins, the $\mathbf{LD_{50}}$

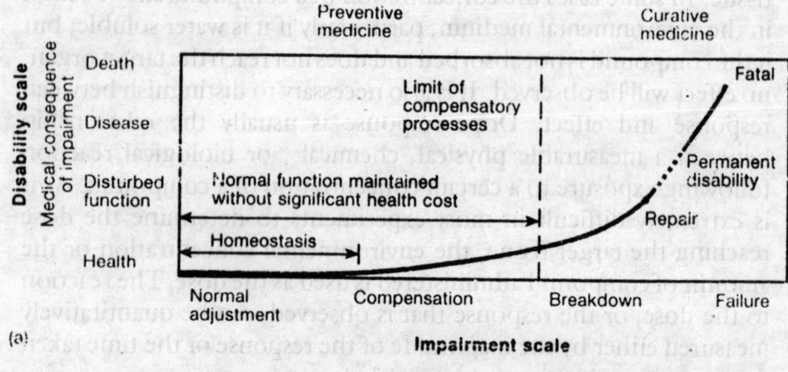

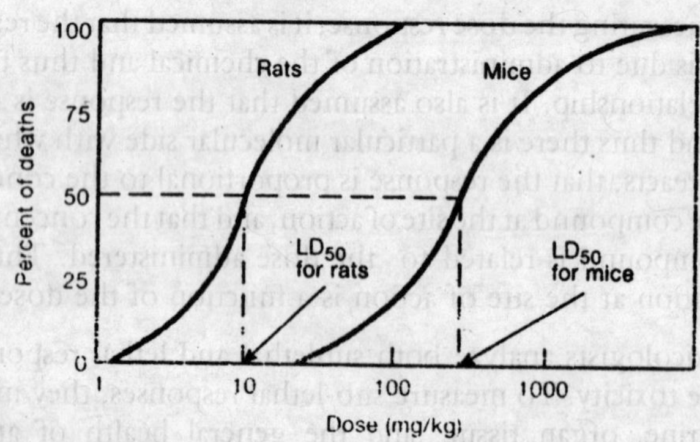

Figure 2.1 : Two types of dose-response curves. (a) Response of an organism as the dose increases, and (b) points where death is the measured response in two different laboratory animals.

value, which is the dose that kills 50% of the individuals in a test population. Table 2.2 shows the wide range in toxicity among a variety of chemicals. Note that LD_{50} values are expressed on a per-body-weight basis, that is the number of milligrams of toxin per kilogram (mg/kg) of body weight lethal to 50% of the test population. (The use of units such as mg/kg is one way of expressing concentration. (See Appendix B.)

Table 2.2 : Approximate Acute LD_{50} of various chemical agents

Agent	LD_{50} (mg/kg)	Approximate Lethal Dose for Humans
ethanol	10,000	between pint and quart
sodium chloride (table salt)	4,000	between ounce and pint
morphine sulfate	1,500	
DDT	100	between teaspoonful and ounce
strychnine sulfate	2	few drops
nicotine	1	
d-tubocurarine (curare)	0.5	one drop
dioxin (TCDD)	0.001	1/500th of a drop
botulinus toxin	0.00001	1/50,000th of a drop

The lethal concentration (LC) for 50% of a population after continuous exposure for 48th is referred to as the 48th LC_{50} or LD_{50} (*LD*=lethal dose). This abbreviation may be altered appropriately to correspond to different exposure times and proportions of the population. Usually the concentration referred to is that in aqueous solution but it may also be a concentration in air. If the test compound is insoluble or sparingly soluble in water, it must be uniformly dispersed for a repeatable LC_{50} to be obtained. If an emulsifier or other solubilizing agent in used for this purpose, it should be chosen carefully to ensure a minimal contribution to the toxicity of the system.

Difficulties arise where organisms are exposed to harmful substances in association with particulate matter, or in their prey or other food. What when is the effective concentration to which they are exposed? The nature of the particulate matter will determine whether the substances are ingested or not. It will also determine the potential for dissociation of toxic substances following ingestion. Normally, only dissociated material can exert toxic effects. In a food organism, toxic substances may be converted to derivatives of greater or less toxicity and may be localized in specific parts of the organisms, some of which may be selectively eaten or discarded by its predators. Therefore, in these cases it is almost impossible to know the effective concentration to which the predator is exposed. The best one can do is to copy the natural situation as closely as possible in the laboratory and express the LC_{50} as a notional concentration, taken as the total available toxic material divided by the weight of particulate matter or by the weight of food organisms. If one knows that a certain derivative of the toxic substance is much more lethal than any of the others, and suitable methods of analysis are available, the derivative concentration in the food organism should be determined since this will be the most significant factors. In medical usage the LC_{50} frequently refers to an injected concentration, and this must be borne in mind when attempting to extrapolate from such studies.

Lethal concentrations have been expressed in a variety of units, most frequently in milligrams per liter of kilogram body weight. Where possible, molarity (moles per liter) or molality (moles per kilogram) should be used as this will give a uniform chemical basis for comparing toxicity. The most information the LC_{50} gives in an idea of the order of magnitude of the lethal under specified conditions. However, it is relatively of the lethal dose under specific conditions. However, it is relatively quick and cheap to determine and provides a basis, however, rough, for initial assessment of the likely hazard from a toxicant and of the effects of various parameters on its toxicity.

Apart from the LC_{50}, a number of other figures may be derived from studies of short term lethality. For example, the minimum lethal dose (*MLD*) is the dose which will kill at least one organism in the test population over the test period. Lucky and Venugopal have proposed that the concept of potential toxicity (p*T*) be introduced on an analogy with pH, i.e. $pT = -\log[T]$ where $[T]$ is the molal concentration of the toxicant and the logarithm is to the base 10. A toxicant with a 24th LD_{50} of 0.001 mol kg would have a 24th $[T_{50}]$ of 10^{-3} and a 24th pT_{50} of 3. The calculated p*T* corresponds directly to the effect to toxicant on the test population. A small pT_{50} should indicate a relatively harmless substance while a large pT_{50} will indicate highly toxic substance, at least in the context of lethality in the test system.

Measurement of short term lethality is one aspect of assessment of acute toxicity. Hunter and Smeets, 1977 have defined acute toxicity as 'the total adverse effects produced by a toxicant when administered a single dose'. And this will be the definition applied here. However, alternative definitions are used, e.g. 'the total adverse effects produced by a toxicant when given in a single dose or multiple doses over periods of 24 hours or less". Therefore, care must be taken to understand the terminology used by a particular author before any appraisal of results can be undertaken. Both definitions quoted above refer total adverse effects which emphasizes the point that there may be many harmful effects before lethality supervenes, and all effects of the toxicant must be monitored throughout testing. Further, through *post mortem* examination must be carried out on any organisms that die. Another parameter may be defined, i.e. the EC_{50} or ED_{50}, which is the concentration or dose which produces a specific ill effect in 50% of the organisms tested. The EC_{50} is subject to the same provisos given for the LC_{50}.

An alternative approach to the measurement of acute toxicity is to measure the ET_{50} which is the median exposure time to a given concentration of the toxic substance required to produce 50% mortality in a test population. This is measured range of concentration and requires continuous monitoring. It therefore involves much more laboratory effort but it can yield more useful information that the 48th LC_{50}. For example, a plot of the ln ET_{50} against ln concentration of toxic substance may indicate a threshold concentration below which the substance is not lethal (Fig. 2.2a). However, it may indicate that the substance is lethal at all concent ations, low concentrations simply requiring long exposures to take effect Fig. 2.2b). Such plot may be strongly suggestive however thev cannot

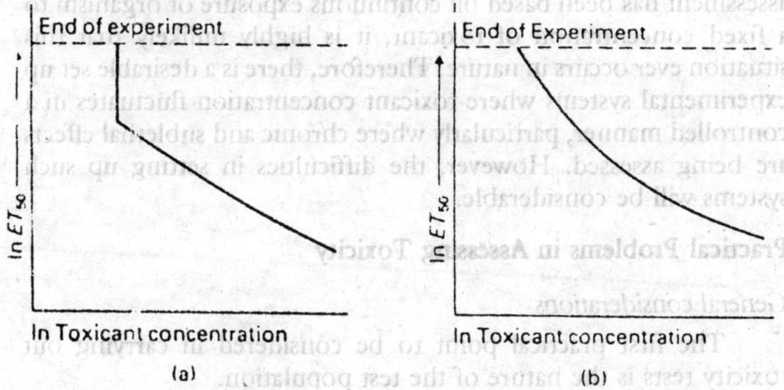

Fig. 2.2 : Possible relationship ET_{50} and toxicant concentration : a graph indicating the existence of a safe threshold concentration : b. graph suggesting that the toxicant will be lethal at all concentrations, given sufficient exposure time.

be regarded as conclusive evidence. Thus, where there appears to be a threshold in these experiments, (which are of fairly short duration) long exposures may lead to significant mortality due to accumulation of the toxic substance. Similarly, an abrupt threshold may appear in a plot which otherwise displays a continuous trend.

Assessment of Long-term Lethality and Chronic Toxicity

In many environment situations the observed problems are caused by the response of organisms to concentrations of toxic substances which are harmful only after long period of continuous exposure. One consequence of such exposure may be death, but the identification of such death as long term lethality from the toxicant is difficult to establish. The death may be due to natural, enhanced or accelerated by the weakened state of the test organism but not directly due to the toxicant. Further, the longer the time over which lethality is determined, the greater the variability of the results. All the limitations of the short-term LC_{50} also apply, so long-term lethality is not normally considered a useful parameter for measurement. Chronic toxicity may be defined as 'total adverse effects produced by a toxicant when administered continuously over a long period of time'. Assessing chronic toxicity is not easy, mainly because of the practical difficulties in maintaining organisms under constant conditions in the laboratory for long durations. The fundamental problems are the same as acute toxicity tests but the time factors greatly exacerbates them. Though almost all laboratory toxicity assessment has been based on continuous exposure of organism to a fixed concentration of toxicant, it is highly unlikely that this situation ever occurs in nature. Therefore, there is a desirable set up experimental systems where toxicant concentration fluctuates in a controlled manner, particularly where chronic and sublethal effects are being assessed. However, the difficulties in setting up such systems will be considerable.

Practical Problems in Assessing Toxicity

General considerations

The first practical point to be considered in carrying out toxicity tests is the nature of the test population.

Ideally, the organisms used should be genetically identical and pathogen free. They should be keep undue sterile conditions in a constant environment and lit with artificial light, as nearly as possible equivalent to sunlight, or a twelve hour day. These are the requirements of the toxicologist whose main concern is a thoroughly repeatably assay of toxicity. Whether there should be prime requirements of an environmental toxicologist is open to question. By definition, the environmental toxicologist is concerned with the list to natural population which are genetically heterogeneous, subject to the effects of pathogens and living in a variable environment. It may be that one important effect of a toxic substance will be to climate selectively sensitive individuals from a population. Another effect may be too eliminate individual affected by a specific pathogen. Other effects may only become apparent under extreme environmental conditions. There are many possibilities, none of which are covered by the analytical toxicologist's ideal population. This is not to say that the carefully selected and maintained populations of organisms, such as inbred rat and mice, which have been developed by medically-orientated analytical toxicologists over the years, do not have their place in environmental toxicology. They are useful for the bio-assay of minute amounts of toxicants, especially where the chemistry of these substances is unknown or does not permit sufficiently sensitive chemical analysis. They may also give pointers to effects on other organisms. Direct extrapolation cannot be justified because of the known magnitude of species and generic differences.

Selecting of organism and their maintenance

In selecting organisms for toxicity assessment, the environmental toxicologist must bear in mind the relevance of the organisms to the environment of interest. The effects of air pollution on plants may be monitored using lichens, which show high sensitivity to many toxicants. Other plants may be used where appropriate, a limiting factors frequently being case of maintenance in the laboratory. This limiting factors is particularly important in selecting animals for testing. Another important limiting factors is ignorance of the normal characteristics and degree of variability of many organisms. There are very few organisms for which there is sufficient fundamental knowledge relating to their life under controlled laboratory conditions, far less under the conditions of their normal habitat.

Assuming that a suitable organisms can be found, i.e. of demonstrable importance in the environment of interest, with a good background of previous study and easily maintained in the laboratory, a breeding population under suitable defined conditions can be established. A breeding population is essential because susceptibility to toxic substances alters with the development stage, typically being greatest in embryonic and larval stages, and in senescence. Suitable defined conditions mean that, until the direct toxicity on the organisms has been assessed, the population should be kept pathogen free as far as possible. Adequate nutrition of known composition must be suppled. In most cases this will be maximal nutrition, with food, intake being controlled only by the organism. Anything less will lead to competition between individuals for the available food resulting in a high degree of variation in food intake throughout the populations. Maximal food intake is not the same as optimal food intake and it is certainly not the most usual occurrence in any natural environment. However, it does provide a defined base from which to work. Other environmental factors to be considered are light, temperature, humidity, day length, nature of living space (which may have behavioural effects), subjection to handling, noise or other disturbances, and, for aquatic organisms, pH, salinity, degree of oxygenation and other parameters of water quality.

Initially, these factors will be established primarily to suit the experimenter's convenience, but the long-term aim must always be to approximate as closely as possible to the particular environment under consideration. This ultimately requires complex, and expensive, control systems and, possibly, an even more complex computer program to analyze the results obtained. At this point, the laboratory experimenter would be wise to consider the possibilities of a properly planned field study. If a suitable area has already been contaminated with the toxic substance of interest, a study in depth should be carried out, monitoring all possible parameters, and the results should be correlated with those from the laboratory.

Lethal levels of air or water pollutant are used by scientists to predict harmful effects to people and the environment. These levels are stated in terms of LC_{50} values, that is, the concentration lethal to 50% of a test organism. In air, the LC_{50} value is expressed as the number of milligrams of substance in a cubic meter (mg/m^3) of air that will kill 50% of a test population when inhaled over a specified period. In water, on the other hand, LC_{50} values refers to the number of milligrams per liter (mg/L), or parts per million (ppm), of a substances required to kill 50% of the population of an aquatic species—usually fish—in a specified time (often ninety-six hours, or four days). For example, the LC_{50} for chlorion dissolved in water is approximately 0.020 ppm for rainbow trout when exposed for ninety-six hours, whereas crayfish can tolerate at least 0.50 ppm. For humans, an exposure of a few minutes to air containing 1,000 ppm chlorine gas is fatal.

While LD_{50} values are important for all of us, the dose-responses effects at lower levels of exposure are of even greater interest because organisms are more likely to encounter these concentration. Although alcohol (ethanol, CH_3CH_2OH) is not usually considered an environmental contaminate. We can use the human response to alcohol to illustrate the gradient of response to sublethal doses.

Alcohol level in humans is measured by analyzing the breath or blood. Blood-alcohol levels are expressed in milligrams of alcohol per 100 milliliters of blood. Drivers in most countries are considered to be intoxicated if their blood alcohol equals or exceeds 100 milligrams per 100 milliliters. This level of intoxication is more commonly expressed as 0.10%. People with lower blood-alcohol levels exhibit some loss of coordination and impaired manual dexterity. An individual's response also depends on the amount of food in the digestive tract, other drugs in the bloodstream, and personal tolerance for alcohol.

Generally, as blood-alcohol levels rise, individuals lose visual acuity, experience a higher pain threshold (course of the expression, "feeling no pain"), and increasingly lose coordination Table 2.3. Relatively high blood-alcohol levels cause vomiting and eventually an intoxicated person slips into unconsciousness. When blood-alcohol levels reach 350 to 400 mg/100 ml (0.35%—0.40%), alcohol functions as an anesthetic. Death ensues at levels only slightly higher than those required to anesthetize a person. Thus, people who become severely intoxicated run the risk of dying from alcohol poisoning.

The toxic response in humans and other animals varies with (1) the sex of the exposed individual, (2) the ability of the individual to develop a tolerance to toxins, and (3) factors interactions, including

Table 2.3 : Effects of alcohol on human behaviour

Blood-Alcohol Level	Effects
0.02%–0.04%	judgement and reasoning somewhat affected; inhibitions lessened
0.05%–0.9%	coordination impaired; judgment and reasoning unreliable
0.10%–0.14%	slow reaction time, blurred vision, lack of coordination
0.16%–0.29%	staggering, slurred speech, visual impairment
0.30%–0.49%	marked lack of coordination, stupor, convulsions
0.50% and higher	coma and death.

synergism and antagonism. Difference in the response of the two sexes to organophosphate pesticides have been shown experimentally. Some of these pesticides are more toxic to female rats and mice, whereas the opposite is true for other organophosphate pesticides. A first-time cigarette smoker typically experiments nausea. Many drug prescriptions warn consumers not to drink alcohol when they use them because of potential adverse interactions.

The very young and very old are generally most sensitive to environmental stresses, including toxins. The young have not yet fully developed their capacity to detoxify toxins, and older individuals lose some of their body's ability to detoxify chemicals. Furthermore, prior to maturity, growing organs are particularly vulnerable to disruption by toxins. The human brain is not fully developed until six or seven years after birth. Thus, children under the age of seven are particularly susceptible to chemicals that adversely affect learning and behaviour. Lead attacks the blood-forming mechanisms, the gastrointestinal tract, and in severe cases, the central nervous system. Lead may also impair the functioning of the heart and the kidneys. In the early 1990s, it was established that lead affects a child's ability to learn. Long-term studies have shown that even children with blood-lead levels not high enough to cause clinical symptoms performed lower than average in schools.

Another complication in assessing the effect of a toxin is that individual species vary in their response to a given toxin. For example, when dioxin, (an extremely toxic chemical) is administered to female rats and female guinea pigs, the guinea pigs die at 1/75 the dose required to kill the rats. When a chemical has widely ranging LD_{50} values for test animals and toxicologists have test results only from animal studies, they use a wider safety margin when setting allowable levels for human exposure to that chemical.

Carcinogens, Mutagens, and Teratogens

Carcinogens are agents that cause cancer. They may be chemical (cigarette smoke), biological (cancer-causing viruses), or physical (ultraviolet light). Cancer is the uncontrolled growth of cells. Each one of us is made up of approximately 100 billion cells, any one of which can be transformed to a maligant (cancerous) cell. Approximately one hundred different types of cancer have been identified.

Carcinogens trigger uncontrolled cell growth in many different ways Fig. 2.3 shows one model of cancer formation. An early first step in the development of cancer occurs when an **initiator**, a type of carcinogen, structurally modifies a gene that normally controls cell growth. (A *gene* is a specific segment of a DNA molecule.) Such altered growth-regulating genes are called **oncogenes**. For a cell to begin to grow uncontrollably, at least two different growth-regulating genes must be altered.

Table 2.4 : Some Environmental Agents that act as Carcinogens, Teratogens, and Mutagens

Carciogens	Teratogens	Mutagens
asbestos fibers	alcohol	benzo(a)pyrene
benzene	arsenic	lead compounds
benzo(a)pyrene	diethylstilbestrol (DES)	mercury compounds
certain chemotherapeutic agents	lead compounds	mustard gas
chromium compounds	lithium	ultraviolet light
diethylstilbestrol (DES)	methyl mercury	x rays
nickel	rubella (German measles)	
nitrosamines	thalidomide	
ultraviolet light	x rays	
vinyl chloride		
x rays		

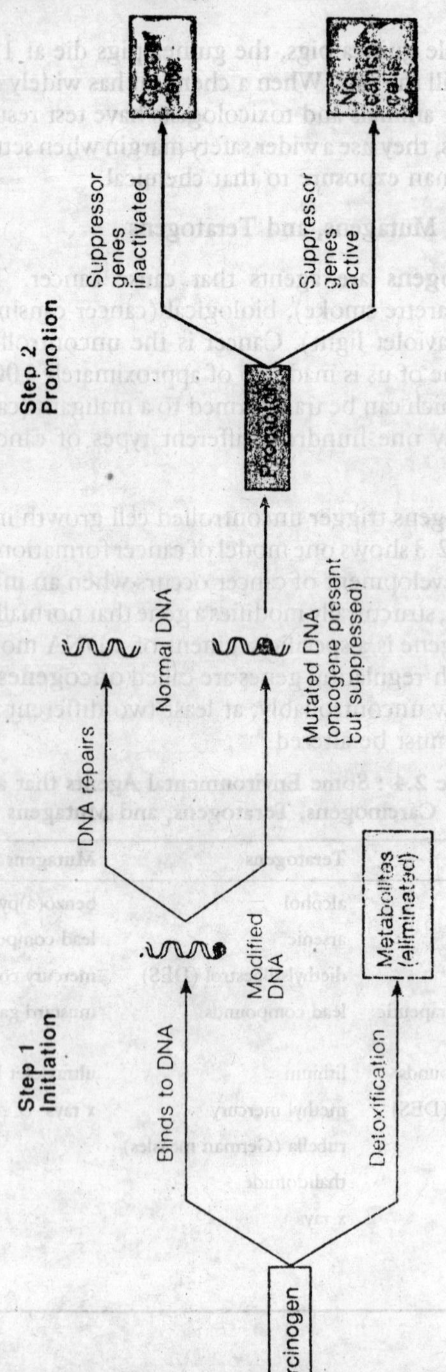

Fig. 2.3 : A probable model for cancer-cell formation.

Formaldehyde, a widely used chemical in industrial glues, is thought to be an initiator. Vapours of formaldehyde trigger the development of malignant tumors in the respiratory tracts of rats. To date, however, formaldehyde has not been shown to cause cancer in humans. In some cases an initiator is a metabolic by-product. While benzo (α) pyrene, a natural product of the incomplete combustion of organic materials, including tobacco, is not an initiator, the body metabolizes it is a related chemical, benzopyrene-7, 8- diol-9, 10 epoxdie, which is an initiator.

Oncogenes usually remain dormant until they are activated by another type of carcinogen, called **promoters**, which may act in several ways. For example, normal cells appears to prevent the activation of an oncogene in an adjacent initiated cell. A substance may act as promoter by killing normal cells that surround an initiated cell. Lifelong exposure to promoters significantly increases the risk of cancer. People whose diets are high in salts or fats, for example, are more likely to develop stomach cancer and colon cancer respectively. Apparently, both dietary factors act as promoters by killing normal cells. Removing promoters from the area of initiated cell that has not yet completed the promotion stage prevents the formation of a cancer cell. Thus, reducing the salts and fat content of one's diet greatly reduces the risk of contracting cancer.

Promoters may also activate oncogenes by inhibiting the action of **suppressor genes**, which prevent oncogenes from initiating uncontrolled cell growth. If suppressor genes are inactivated, oncogenes can then spur tumor formation. A better understanding of tumor suppressor genes may well provide the means for future anti-cancer therapy based on nature's own method of protecting against cancer.

For most toxins, a threshold dose is necessary before a response occurs. Many scientists believe, however, that exposure to any amount of a carcinogen poses some risk; that is, there is not margin of safety. This property of carcinogens poses problems for scientists and public policy makers. For example, the chlorine added to public water supplies kills most pathogens, but its use also results in the formation of small amounts of carcinogens such as chloroform ($CHCl_3$). Because it is impossible to reduce chloroform levels in drinking water to zero, public policy regulators set the allowable concentration of chloroform and other potential carcinogens at a level that will cause one additional case of cancer in a population of one million people. Thus, they opt for a minimal risk for contracting cancer (with chlorination) rather than the greater risk of contracting a water-borne disease (without chlorination)..

The incidence of cancer (number of cases per million population) is dose-dependent, while the severity of the response (cancer) is independent of dose. This means that as a population is exposed to higher levels of a carcinogen, we would expect more cases of cancer to develop. But once an individuals has contracted cancer, the dose of the carcinogenic agent has little to do with the severity of the disease. This distinction between toxins and carcinogens is the reasons that exposure regulations for substances classified as carcinogens are much stringent than for toxins.

Mutagenesis is the creation of mutations by chemicals or ionizing radiation which bring about alteration to DNA to produce inheritable traits. Although mutations can occur naturally in the absence of xenobiotic substances, most mutations are harmful.

Mutagens cause mutations, which are inheritable changes in the DNA sequences of chromosomes. All organisms have a package of chromosomes that are the result of evolution over numerous generations. Hence, most organisms are well adapted for survival in their particular habitat. Mutations are often harmful because they involved a random change in the natural functioning of chromosomes or their component genes; such changes (mutations) do not usually benefit the organisms's offspring.

Some common disease caused by mutations include cystic fibrosis, sickle-cell anemia, hemophilia, and Huntington's disease (a loss of neural muscular control in middle -aged adults). People who have a family history of these or other inheritable diseases are advised a seek genetic counseling before having children.

Teratogenesis is the creation of birth defect arising from damage to embryonic or foetal cells or from mutations in egg or sperm cells. The biochemical mechanisms of teratogenesis include: enzyme inhibition by xenobiotics, deprivation of essential nutrients and alteration of the placental membrane.

Teratogens cause abnormalities in a developing fetus. People frequently associate teratogens with severe deformities, such as the shortening or absence of arms or legs. Such was the impact of thalidomide, a sedative that was taken by thousands of pregnant women in West Germany during the early 1960s. Sometimes the effect of a teratogen does not appear until many years after the mother has been exposed.

NATURAL DEFENCE MECHANISMS: DETOXIFICATION ROUTES

The body was a number of **detoxification mechanisms**, that is, means whereby the body metabolizes or excretes toxins.

Ingested toxins enter the body via the gastrointestinal tract, a long, highly specialized tube that extends throughout the length of the body's trunk (Fig. 2.4). Ingested materials are processed in the mouth, stomach, small intestine, and large intestine. Glands in the

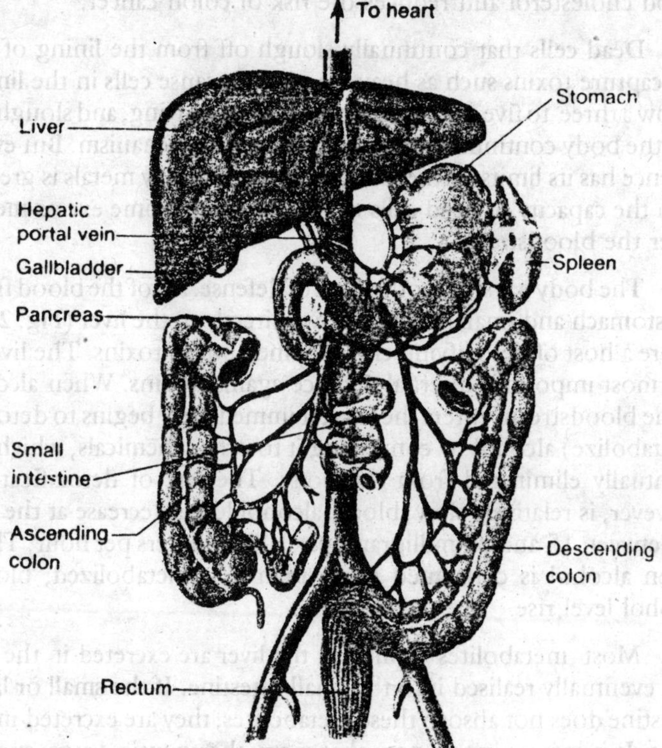

Fig. 2.4 : Blood leaving the digestive tract flows through the hepatic portal vein to the liver before being circulated to other parts of the body.

lining of the stomach and small intestine together with the liver and pancreas, release fluids and enzymes essential for digestion. These fluids and enzymes convert the complex organic molecules of our food into simpler organize molecules that are absorbed (mainly by the lining of the small intestine) and taken up by the bloodstream.

The gastrointestinal fact has a variety of defence mechanisms against toxins. The same processes that digest food also metabolize ingested toxins into breakdown products called **metabolites**, most of which are not toxic. In addition, certain materials in the gut capture toxins and prevent them from moving into the bloodstream. For example, soluble fiber tends to capture chemicals such as cholesterol and carry them out of the body in feces. For this reason, high-fiber diet (a diet rich in grains, fruits, and vegetables) lowers blood cholesterol and reduces the risk of colon cancer.

Dead cells that continually slough off from the lining of the gut capture toxins such as heavy metals. Because cells in the lining follow a three-to five-day sequence of dividing, dying, and sloughing off, the body continually renews this defence mechanism. But every defence has its limits. If the concentration of heavy metals is greater than the capacity of dead cells to remove them, some excess metals enter the bloodstream.

The body has still another line of defense. All of the blood from the stomach and small intestine flows directly to the liver (Fig. 2.4), where a host of detoxifying enzymes metabolize toxins. The liver is our most important internal defence against toxins. When alcohol in the bloodstream enters the liver, it immediately begins to detoxify (metabolize) alcohol by converting it to other chemicals, which are eventually eliminated from the body. The rate of detoxification, however, is relatively slow; blood-alcohol levels decrease at the rate of between 15 and 20 milligrams per 100 milliliters per hour. Thus, when alcohol is consumed faster than it is metabolized, blood-alcohol level rise.

Most metabolites formed in the liver are excreted in the bile and eventually realised into the small intestine. If the small or large intestine does not absorb these metabolites, they are excreted in the feces. It is interesting to note that many therapeutic drugs, such as those used to treat cancer, are administered intravenously (through the veins), rather than orally. Thus, the drug is not metabolized by the live before it reaches its target organ.

Enzymatic breakdown of toxins does not always produce a metabolite that is less toxic than the parent substance. The detoxification of two common alcohols, methanol (wood alcohol, CH_3OH) and ethanol (grain alcohol, CH_3CH_2OH) illustrate the differing toxicity of metabolites. Both methanol and enthanol are readily absorbed by the lining of the stomach and taken up by the bloodstream. In the liver, the same enzyme system breaks down the two alcohols, as illustrated in Figure 2.5. The metabolites of ethanol, acetic acid and acetaldehyde, are relatively non-toxic, but the metabolite of methanol, formic acid, travels, through the bloodstream and damages the light-sensitive region of the eye. Permanent blindness results from the ingestion of too much methanol. Wood alcohol is an example of a chemical that does not have to be directly toxic to cause harm; sometimes the enzymatic breakdown of a chemical to a more toxic metabolite is the actual cause of toxicity.

In the long run, even the liver has a limited ability to cope with continual exposure to toxins. A chronic abuser of alcohol, for example, eventually develops fibrous tissue in the liver, which constricts the flow of blood. Restriction of blood flow reduces the capacity of the liver to detoxify alcohol. Such degeneration of liver function, called *cirrhosis*, is eventually fatal. Although the liver is the principal defence against toxins, the kidneys also lay a significant role. They filter the blood and subsequently excrete toxins from the body in urine.

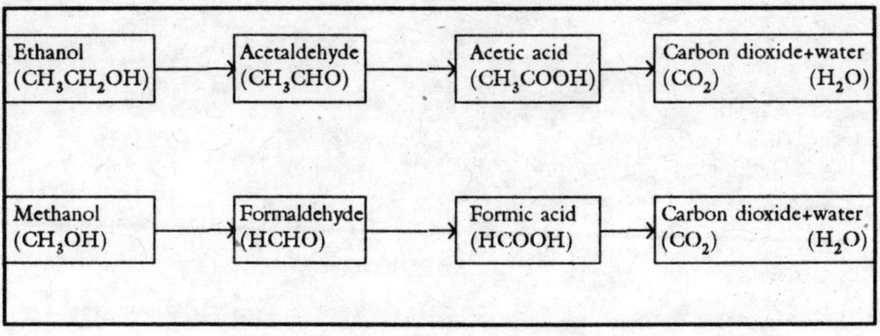

Fig. 2.5 : The metabolites formed when two common alcohols are ingested and metabolized

REFERENCES

Anderson J.E. (1974) *Environment*, 16(3), 6-11.

Hunter, W.J. and J.G.P.M. Smeeth (1977). *The evaluation of Toxicological Data for the Protection of Public Health*, Pergamon press, Oxford.

Manahan, S.E. (1991) *Environmental Chemistry* (5th Edn.) Lewis Publishers, Chelsea, Mich.

Lucky, T.D. and B. Venugopal (1977) *Metal Toxicity in Mammals*, Vol. 1, Plenum Press New York.

Turk, J. and A. Turk (1984) *Environmental Science*, Saunders College Pub. Philadelphia.

3

Toxicology of Gaseous Pollutants

The country dweller entering a large town can usually detect, by the nose, the eye, or the sensation of the skin, a definite deterioration in the quality of the air. The deterioration is more noticeable in some towns than others; but even when town air is at its best the country air is better still. It used to be commonly believed that town air is stale by reason of having been breathed by so many people, yet in the most densely populated areas the hourly breath requirements of the people would be satisfied by a layer of air less than 1 cm thick. The only detectable chemical differences between town and country air are differences due to atmospheric pollution.

The composition of air is given in Table 3.1 Among the major constituents, water vapour is extremely variable, its amount depending strongly on temperature and relative humidity. The concentrations of the major constituents can also be expressed as per cent by volume, and the major constituents in units that are 10,000 times smaller : parts per million (ppm).

The nature and degree of air pollution have varied over time and the attitude of man towards it has also changed. As the human population grew, so did the levels of environmental pollution. Air

Table 3.1 : Approximate Composition of the Dry Atmosphere

(i) Major Constituents			(ii) Minor Constituents		
	g m–3	% (by volume)		mg m–3	ppm
Air	1184	100	Neon	14.9	18
Nitrogen	895	75	Helium	0.85	5.2
Oxygen	274	23	Methane	0.79	1.2
Argon	15.2	1.26	Krypton	3.4	1.0
Water vapour	var	0.70	Hydrogen	0.04	0.5
Carbon dioxide	0.57	0.04	Nitrous oxide	0.90	0.5
			Xenon	0.43	0.08

pollution was not considered a major health problem up to the end of the 13th century. By that time coal was used extensively, and gases emitted from the burning became the most important air pollutant. As early as 1900 the combination of coal smoke and fog, commonly known as *smog*, became associated with significant increases in urban death rates. The potential hazards associated with the exhaust gases of the internal combustion engine were reported as early as 1915. In the past few years substantial evidence has been accumulated on the adverse affects of air pollution on human health and well being.

Air pollution occurs when either chemical or physical agents are present in the air in quantities that can produce significant effects on either man, animals, vegetation, or abiotic materials. These substances may be emitted from certain man-made sources and from natural events. Natural air pollutants includes forest fires, bacteria, pollen, and dust particles. Sources of man-made pollutants are emissions from transportation, industrial factories, electric power plants, coal and oil combustion, and the burning of waste. The number of chemical emitted into the atmosphere from these sources is very large and may be different from one area of the country to another.

There are two groups of air pollutants are divided: those emitted directly into the atmosphere from specific sources (primary pollutants) and those that results from the interaction between two or more primary pollutants (Secondary pollutants) through complex chemical reactions that occur in the atmosphere under certain conditions.

The primary pollutants include organic compounds, compounds containing nitrogen, sulphur, carbon, and halogens, radioactive compounds, and particles ranging in size from 0.1 to 100 μm in size.

The organic compounds includes saturated and unsaturated aliphatic and aromatic hydrocarbons together with a variety of oxygenated and halogenated derivatives. These pollutants may be emitted as vapors, liquid droplets, or solid particles. Outdoor stationary sources such as chemical plants and petroleum refineries are the largest contributor of these pollutants. Individuals may also be exposed to significant amounts of volatile organic compounds such as benzene and other organic solvents.

CARBON MONOXIDE

Carbon monoxide is a colourless, odourless gas produced by the incomplete combustion of carbon-containing fuels and by some biological and industrial processes. Every year human activities may put some 1500 teragrams (1 Tg= 10^{12}g) of carbon monoxide into the Earth's atmosphere compared with the 1200 Tg/year from natural sources.

The major source of carbon monoxide emissions at breathing level outdoors is the exhaust of petrol-powered motor vehicles, the diesel engine) compression ignition), when properly adjusted, emits little carbon monoxide. Locally, high concentration of carbon monoxide may occur near industrial plants such as power stations, petroleum refineries, iron foundries and steel mills as well as in the vicinity of refuse burning, whether in incinerators or openly. Carbon monoxide concentrations in urban areas are closely related both to motor-traffic density and weather conditions. Carbon monoxide levels show a distinct diurnal pattern with peaks corresponding to the morning and evening traffic rush-hours. Levels of this pollutant decrease very rapidly with distance from emission sources. Short-term concentrations of even higher concentrations have been observed in confined spaces such as tunnels, garages, loading bays, underpasses underground car parks, and in narrow congested roadways. At normally encountered levels, there are no known adverse effects of carbon monoxide on vegetation and materials.

TOXICITY PROFILE

Health Effects

Carbon monoxide is absorbed through the lungs and reacts with haemoproteins, especially with haemoglobin of the blood. This in turn results in a reduction of the oxygen carrying-capacity of the blood, and also interferences with the release of the oxygen which is carried to the tissues. Carbon monoxide has an affinity for haemoglobin that is 200–400 times greater than that of oxygen, and carboxyhaemoglobin (COHb) is therefore a more stable compound than oxyhaemoglobin. Carbon monoxide is not a cumulative poison, but is excreted or absorbed, depending upon the level of carbon monoxide in the ambient air, the amount of carboxyhaemoglobin in the blood, barometric pressure, the duration of the exposure, and the rate of ventilation of the lungs. A useful approximate relationship relating ambient carbon monoxide concentration to carbon monoxide saturation of the blood is that for every 1 ppm of carbon monoxide with which the body is in equilibrium (by diffusion through the alveolar sacs), 0.165 per cent of the body's haemoglobin will be combined in the form of carboxyhaemoglobin.

By far the most common cause of high carboxyhaemoglobin concentration in people is the smoking of tobacco and inhalation of the products by the smoker. Typically, cigarette smokers generally have a mean carboxyhaemoglobin level of 5 per cent compared with 1 per cent in non-smokers. However, traffic policemen, garage attendants, and drivers of taxis and trucks exposed to vehicle exhaust emissions experience increase of carboxyhaemoglobin levels up to about 3 per cent. There is evidence that a carboxyhaemoglobin level of between 1 and 2 per cent affects behavioural performance and can aggravate symptoms in patients with cardiovascular disease; a level of between 2 and 5 per cent causes impairment of vigilance and time-interval discrimination, visual acuity, brightness discrimination, and certain other psychomotor functions; while a level exceeding 5 per cent is associated with cardiac and pulmonary functional changes. As levels increase above approximately 10 per cent, the adverse effect include headaches, fatigue, drowsiness, reduced work capacity, coma respiratory failure and ultimately death. Given such effects it seems desirable to keep carboxyhaemoglobin levels generally below 2 per cent even though it is recognized that with carbon monoxide interfering with oxygen-carrying to crucial tissues such as the brain, heart and muscles, there may be not threshold level for health effects.

The World Health Organization (1972) has suggested a long-term goal of 10mg/m^3 (9ppm) for an l8-h period and 40 mg/m^3 (35 ppm) for a 1-h averaging period. The 8-h averaging is employed because it takes from four to twelve hours for the carboxyhaemoglobin level in the human body to reach equilibrium with the ambient carbon monoxide concentration.

The effects of CO inhalation have been extensively documented. A unique feature of CO exposure is the biological marker of the dose that an individual has received. By measuring the blood level of COHb the CO exposure can be calculated. In as much as the major effect of CO on the body is related to the capacity for transporting O_2 to the tissue, all organs that require a continual high O_2 supply for normal functions are critical targets. The most important of these the heart and the central nervous systems (CNC).

The basic mechanism governing the interaction of CO and O_2 with hemoglobin indicates that CO combines with the reduced hemoglobin molecule much more rapidly than does O_2, thus decreasing the O_2 transport capability of the blood. The tenacity of binding between CO and hemoglobin is more than 200 times that of O_2. Thus, a small amount of inhaled CO can usurp a large proportion of hemoglobin in the blood, seriously impairing the transport of O_2 to extrapulmonary tissue. The alveolar CO levels can be crudely converted to COHb by dividing the CO in the air ppm by 7. The half-life of COHb in the blood is about 4 h.

Acute exposure to CO levels of 200 to 1200 ppm can cause headaches at 10 to 20% blood COHb levels, mental confusion and incoordination at 40% convulsions at 50 to 60%, and death at COHb levels about 70%. Normal recovery from a coma can occur if tissue anoxia is not too severe. Adaptation to chronic levels of CO can occur through an increased hematocrit, increased hemoglobin, and increased blood volume.

Since the heart requires a continuous supply of 02 to maintain electrical and contractile integrity, the critical important of the cardiovascular system during exposure to CO is evident. During periods of tissue hypoxia, the heart compensates by increasing both its rate and its output to meet normal O_2 demands of the body. Alteration of electrical activity and decline in contractile force and ventricular fibrillation soon follow the induction of myocardial hypoxia. The threshold for this latter effect seems to be about 100 ppm CO (9% COHb). The most susceptible individuals at risk are those with clinical heart disease or chronic obstructive pulmonary disease.

Students have shown that exposure to 50 ppm CO can cause chest pain in patients with angina. These individuals had a COHb level of 2 to 4.5%. Animal studies have indicates that CO exposure may promote atherosclerosis, especially in individuals on high fact diets.

Exercise capacity is reduced with CO exposure. When the COHb exceeds 45%, human subjects are unable to perform slight physical exertion.

COHb levels of 7% have been associated with decrements in attention span and learning ability in humans. Similar effects have been observed in animals exposed to CO.

While adults may have a certain capacity to adjust to moderately high COHb levels without irreversible effects the fetus is much more susceptible to decreases in tissue oxygen supply to even a moderate CO exposure could have a deleterious effect during fetal development. Exposure could have a deleterious effect during fetal development. Exposure of pregnant rabbits to CO (10% COHb) resulted in lower birth weights and increases in neonatal mortality and limb deformities. The offspring of female rats exposed to CO (16% COHb) during gestation had significant learning deficits, indicating the susceptibility of the fetus to CO.

OXIDES OF NITROGEN

Oxides of nitrogen (NO_X) are produced by natural processes, including bacterial action in the soil, lightening and volcanic eruptions and by human activity during combustion processes at temperatures higher than about 1000°C. Nitric oxide (NO) and nitrogen dioxide (NO_2) are the most important oxides are nitrogen for pollutions studies because other oxides of nitrogen such as nitrous oxide (N_2O), denitrogen trioxide (N_2O_3), denitrogen tetroxide (N_2O_4), denitrogen pentoxide (N_2O_5), and nitric acid (HNO_3)

vapour, which may exists in ambient air, are not known to have any biological significance. The principal emission of oxides of nitrogen from human activities are from the combustion of fossil fuels in stationary sources (heating, power generation) and in motor vehicles (internal combustion engine). The relative contribution of each source varies from country to country in relation to differences in fuel use.

Toxicity Profile

Of the oxides of nitrogen , nitrogen oxide (NO_2) is the most toxic. Exposure to NO_2 can lead to wide variety of respiratory effects in both man and animals. Like ozone, both reversible and irreversible effects may be caused and the responses depend on the concentration, duration, specific pattern of the exposure, and the species studied. While the magnitude of the effect increases with both concentration and time, concentration has the most pronounced effect.

Annual mean concentrations of nitrogen dioxide, monitored as the indicators of levels of gaseous oxides of nitrogen, in urban areas throughout the world are typically in the range of 20–90 μm/m^3 (0.01-0.05 ppm) with maximum 24-h being typically between two and five times greater than annual means and maximum hourly means between five and ten times greater than annual means. However, in some areas, such as near industrial plants producing nitric acid or explosives, or near power stations, very high nitrogen dioxide levels may occur. Exceptionally high nitrogen dioxide concentrations may also occur indoors from sources such as gas-fired heaters, boilers and cookers as well as from cigarette smoking.

NO_2 reacts with the high humidity and temperature of the respiratory tract resulting in formation of chemicals such as HNO_3 and HNO_3. Like Ozone, NO_2 has strong oxidative potential. Nitrate have been detected in the blood and urine of animals exposed to NO_2 indicating that NO_2 reacts with various cellular constituents producing nitrates.

Experimental studies with man have indicated that the odour thresholds are about 0.1 ppm. The perception of odour is enhanced by high humidity and is attenuated by increased duration of the exposure.

Clinical studies indicate that the healthy subjects at rest to do now show difficulties in breathing at concentration $<.2.0$ ppm suggesting that acute exposure to outdoor ambient levels of NO_2 may not acutely impair healthy individuals. Individuals with chronic lung disease had an adverse response after a short exposure to 0.3 ppm.

When exercising, patient with asthma may be more sensitive to NO_2 exposure. The airway passages, may be constricted, reducing their ability to draw air into the lungs.

Epidemiological studies have not been able to show clearly a significant difference in lung function individuals exposed within the community. The studies indicate possible diminished lung function and increased in both lower and upper respiratory disease in school children at ambient concentrations.

NO exposure cause numerous morphological effects including hypertrophy/hyperplasia and an increase in blood cells in the upper respiratory tract. At low concentration these changes will resolve but more permanent damage occurs at concentrations > 0.2 ppm.

NO_2, or its by-products, produces several systemic responses: (1) reduction in body weight, (2) changes in white and red blood cells, platelet, and hemoglobin (3) altered clinical chemistries, (4) cardiovascular system effects, (5) hepatic changes and (6) effects on the kidney and urine contents.

Effects on vegetation

Oxides of nitrogen rank second to sulphur compounds in their contribution to acid rain which may affect terrestrial and aquatic ecosystem. In northeastern United States, for example, 30 per cent of the acidity of precipitation (below pH 5.6) is believed to be caused by the oxides of nitrogen producing nitric acid, 65 per cent to sulphuric acid and 5 per cent to hydrochloric acid. However, whereas the contribution of sulphate to the problem of aid precipitation is leveling off, that of nitrate is increasing. In addition to the indirect effects of oxides of nitrogen via acid precipitation on vegetation, prolonged exposure to nitrogen dioxide concentrations of 470–1880 $\mu m/m^3$ (0.25-1.0) may suppress growth of such plants as tomatoes, pinto beans and navel oranges.

Atmospheric Effects

Significant atmospheric effects of oxides of nitrogen include their role in reducing visibility and their potential for causing a global surface temperature increase. In the atmosphere, nitric, oxide emissions are readily converted to nitrogen dioxide and nitrate aerosols, both of which reduce visibility. Nitrogen dioxide absorbs visible light (and strongly absorbs ultraviolet radiation), and at a concentration of 470 $\mu g/m^3$ (0.25ppm) will cause an appreciable reduction in visibility. Generally, nitrogen dioxide appears to be less important than nitrates in causing visibility reductions, while nitrate currently appear to be less important than sulphates because of lower emissions and because of the variable size range of nitrates.

If the global atmospheric concentration of nitrous oxide increases, this may lead to a surface warming because of 'greenhouse effect'— that is, the ability of nitrous oxide to absorb infra-red radiation. Nitrous oxide is mostly maintained at its present concentration by biological decay and conversion processes taking place in soil and in the oceans-processes referred to as 'de-nitrification'. It has been suggested that the increasing use of nitrate fertilizers by humankind may accelerate the biological production of nitrous oxide and raise its atmospheric concentration. An increase of nitrous oxide concentration in atmosphere may also lead to a decrease in the stratospheric ozone concentration which may also affect surface temperatures.

PHOTOCHEMICAL OXIDANTS

Photochemical oxidants are secondary pollutants produced by the action by sunlight on an atmosphere containing reactive hydrocarbons and oxides of nitrogen. The complex series of photochemical reactions produces various oxidants with the most important being ozone and peroxyacetyl nitrate (PAN) as shown in Fig. 3.1. Regular measurements of ambient photochemical oxidant concentrations are usually confined to ozone because it is the most plentiful oxidant in a polluted urban atmosphere and easiest to detect-though probably not the most toxic.

Ozone can form naturally in the atmosphere such that background mean monthly concentrations vary from 0.005 to 0.04 ppm by volume (10–80 $\mu g/m^3$) depending upon latitude and month of

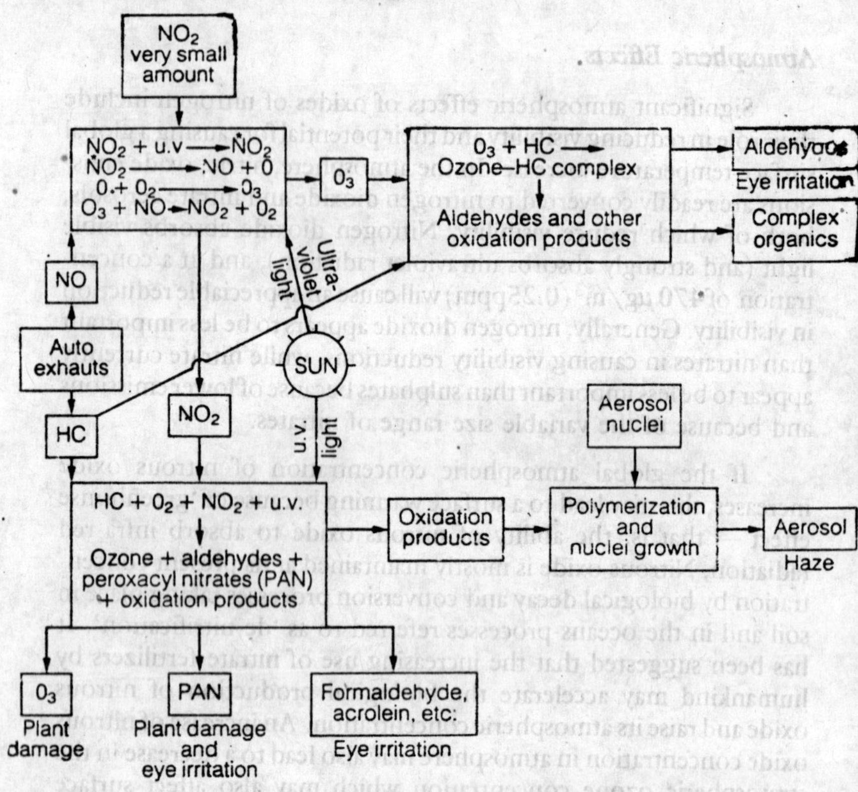

Fig. 3.1 : Photochemical smog formation

the year. Hourly background values ranges from 0.005 to 0.05 ppm (10–100$\mu g/m^3$). In contrast, ozone levels in urban areas may reach peak hourly concentrations of 0.15 to 0.40 ppm (300-800$\mu g/m^2$). In some large cities, maximum one hour oxidant concentrations exceed 0.1 ppm (200$\mu g/m^3$) on 5-30 per cent of days. In 1985, 0.10 ppm of ozone was exceeded on 107 days of the year in Los Angeles, on 173 days in Pasadena, and on 166 days in Azusa (US SCAQMD, 1986).

Polluted air masses from urban and industrial areas can affect suburban and rural areas in the direction of the prevailing wind for considerable distances.

Toxicity Profile.

An array of health effects is caused by exposure to ozone.

Although the precise mechanisms of action of ozone at the molecular level is still unknown, it is agreed that ozone toxicity is related to its oxidative capacity. Oxidants can react with virtually every class of biological substance. It is likely that the actual mechanisms of ozone toxicity may involve a combination of several interacting mechanisms; however, it appears that the specific site of injury is the cell membrane. Thus, the toxic effect expected would be an alteration in the cell permeability and leakage of essential intracellular enzymes from the cell. If the exposure is severe, the damaged cell may not recover.

The magnitude of ozone health effects if highly dependent on concentration, duration of exposure, and the activity level of the subject. Pulmonary functional change and symptoms of toxicity have been measured in both clinical and epidemiological studies. Within the normal population some subgroups respond differently. Differences in response may be due to age, race, gender, disease status, and the specific endpoint measured. Exposed adults report symptoms including chest tightness, irritation, and cough. Exposure can cause decrement in pulmonary function including rapid, shallow breathing, in the ability to inspire maximally, and decrements in forced expiratory function.

Health Effects

The odour (like that of weak chlorine) threshold for ozone is approximately 0.008-0.02 ppm (15–40 $\mu g/m^3$) but the threshold for adverse health effects on the population is about 0.10 ppm (200 $\mu g/m^3$) measured as a one-hour average concentration. Eye irritation occurs at about 0.10-0.15 ppm (200–300 $\mu g/m^3$) with the intensity of eye irritation increasing progressively as concentrations exceeds this value. However, other oxidants such as PAN, peroxybenzol nitrate (PBN) and acrolein are even stronger eye-irritating oxidants than ozone. The threshold levels, determined as maximum hourly concentrations, at Los Angeles in 1974 were 0.05 ppm (100$\mu g/m^3$) for causing headaches, 0.15 ppm (300 $\mu g/m^3$) for eye irritation, 0.27 ppm (530 $\mu g/m^3$) for coughs, and 0.29 ppm (580 $\mu g/m^3$) for chest discomfort.

Epidemiological studies have discovered no increase in mortality with increase in oxidant levels. (Only when high temperature and high oxidant concentrations coincide does daily mortality of the aged increase, and temperature alone is known to cause deaths by hyperthemia. With regard to the effects of ozone-one respiratory functions, much of the evidence has been provided by controlled laboratory experiments. Numerous experiments with healthy male subjects exposed to ozone concentrations ranging from 0.1 and 1.0 ppm (200–2000 $\mu g/m^3$) have shown that increased airway resistance and decreased ventilatory performance occurs although there was great variation individual response.

Numerous other types of responses (i.e., structural, biochemical, and physiological) have been identified in animal studies. Acute exposure to ozone concentration < 1.0 ppm produces evidence of significant structural changes in cells that line the respiratory tract and the alveoli. With chronic exposure further structural changes may occur which can develop into chronic obstructive lung disease (i.e. fibrosis) which is not reversible.

Changes in various biochemical parameters have been used as markers of pulmonary response to ozone.

Several host defense systems are normally operative in protecting the body from infectious and neoplastic diseases. In the upper respiratory tract, ozone reduces the effectiveness of the mucociliary escalator by increasing mucus secretions and reducing cilia beating activity.

Within the gaseous exchange region of the lung the first line of defense is the alveolar macrophage. O_3 reduces the ability of these cells to engulf (phagocytize) microorganisms and non-viable particles deposited in the lung reduce their ability to migrate disrupt macrophage membrane integrity and reduce lysosomal enzyme and bactericidal activity. Numerous investigators have reported that these effects cause increases in the incidence of respiratory infections.

Changes in the central nervous system and behavioral pattern in animals at concentrations less than 1.0 ppm have been reported. Changes have been detected in the serum of animals exposed to O_3 indicating increases in cholesterol and decreases in total lipoprotein triglycerides.

Pregnant animals and developing fetuses may be at risk from photochemical oxidants. Effects reported have included decreased average maternal weight gain, increased fetal resorption rate, and delays in certain behavioral responses.

Effects on Materials and Vegetation

Ozone, an oxidant much stronger than oxygen, causes cracking of stretched rubber at hourly concentration of only 0.01–0.02 (20-40 $\mu g/m^3$), although ozone inhibitors can be build into rubber products such as vehicle tyres and rubber insulation. Ozone also attacks the cellulose in textiles, reducing the strength of such items and all oxidants cause some fading of fabrics and dyes. Textile fabric affected include cotton, acetate, nylon and polyester. Erosion of exterior painted surfaces due to attack on the organic binder in the paint also occurs. Even inside buildings such as museums and art galleries, where ozone levels may be half of those found outside, there is concern that the organic materials in antiquities might be damaged.

Oxidants cause acute and chronic injury to plants, causing necrotic patterns on levels, growth alterations, reduced yields and reductions in the quality of the plant products. Typical effect of ozone are stippling or flecking (brown sports or flecks which subsequently turn white) on the upper surface of leaves. Vast areas of trees are dying as the result of prolonged exposure to photochemical oxidants. The tree are injured by the high ozone concentration and become very susceptible to fatal pest outbreaks of pine-bark beetles. One the western slopes of the southern Sierra Nevada Mountains, where ozone levels are much lower, widespread injury by chlorotic mottle on the needles and more advanced stages of chlorotic decline have been attributed to the smog. Similar damage has been observed in other regions of the United states, in Mexico and ln Israel, while in many parts of Europe die-back of forests appears to the happening on a massive scale. Air pollutants, including ozone, are considering to be largely responsible for the damage to European forests.

It is estimated that each year there are huge direct commercial losses from ozone damage in the United States. For California alone, total crop losses are estimated to reach about one billion dollars a year. Other crops especially sensitive to ozone damage include spinach, tomatoes, pinto beans and tobacco. The tobaçco (*Nicotiana tabacum*) strain Bel-W_3 may be employed as a bioindicator of ozone levels in excess of 0.05 ppm (100 $\mu g/m^3$) as this is the level at which necrotic damage to the leaves occurs in this species. In general, an ozone concentration of 0.05 ppm (10 $\mu g/m^3$) for a four hour period or 0.03 ppm (60 $\mu g/m^3$) for an eight-hour period appears to be the threshold of damage for sensitive plants. However, if even low concentrations of sulphur dioxide are present, the time for the effects to occur is reduced. In addition, 0.01 ppm (20 $\mu g/m^3$) of PAN for a six-hour exposure or 0.015 ppm (30 $\mu g/m^3$) of PAN for a four-hour exposure has been observed to cause under surface glazing, silvering or bronzing of new leaves of sensitive plants.

Atmospheric Effects

Secondary pollutants formed during the chemical reactions that create photochemical pollution episodes cause a marked reduction in visibility-hence the use of the term 'smog'. The smog varies in colour according to the nature of the various constituents of the smog, being brownish in Los Angeles, where as the London smogs of 1976 were of a greenish hue. Ozone also plays a part in producing regional hazes by promoting the oxidation of sulphur dioxide to sulphate particles. Ozone itself is virtually a colourless gas (very faintly blue).

Increased anthropogenic emission of the precursors of ozone formation have led to increasing tropospheric ozone concentrations in the middle latitudes of Northern hemisphere.

Suspended Particulate Matter and Sulphur Dioxide

The term 'suspended particulate matter' refers to the wide range of finely divided solids or liquids disperse into the air from combustion processes (heating and power generation), industrial activities, and natural sources. Suspended particulars range in size from 0.1 up to about 25 μm in diameter.* The constituents of suspended particulate matter vary over time and space, although typical constituents in urban areas include carbon or higher hydrocarbons formed by incomplete combustion of hydrocarbon fuels. Up to 20 per cent of the total suspended matter may consist of sulphuric acid and other sulphates (as much as 80 per cent of the particular less than 1 μm in diameter).

A major problem studying the health effects of suspended

* One micrometre (1 μm) equals 10–6 metre. The term micron (μ) is the non-SI name given to the micrometre.

particular matter is that there are two differing techniques widely employed to monitor has pollutant. Whereas in Europe, measurements of suspended particulate matter are based on soiling properties,* in the USA the monitoring technique is based on weight.** The World Health Organization (1976a) recommends the former suspended particulate matter sample to be referred to as "smoke" (or sometimes 'soot'), and the latter as 'total suspended particulate' (TSP). Although comparative Sulphur dioxide (SO_2), a colourless gas, is the principle form of sulphur oxide omitted from the combustion of coal and oil. In the ambient air SO_2 may be inhaled either as a gas or bound to airborne particulates. Depending on their physical and chemical properties, these inhaled substances are deposited in various regions of the respiratory tract where several chemical reactions may take place. These reactions include oxidation of SO_2 to sulphate which is excreted in the urine, formation of sulphurous acid which may be absorbed into the bloodstream through the pulmonary capillaries, reversible binding to proteins, or irreversible autoxidation with the formation of free radicals. These chemical reactions may be expected to alter the toxicity of SO_2. Due to its high water solubility SO_2 is deposited primarily in the upper respiratory tract where the relative humidity is high; however, SO_2 can penetrate more deeply into the lung during mouth breathing and with exercise. Because SO_2 often co-exists with airborne particulate

---

* The smoke-shade or reflectance method draws air though a filter-paper such that smoke particles suspended in the air are retained on the paper, forming a stain. 'Smoke' is considered to include particles of approximately 10 μm diameter or less. The density of the stain depends partly on the mass of smoke particles collected and partly on the nature of the smoke. For example, the blackness per unit mass of diesel particulate matter is three times that of smoke from bituminous coal combustion and seven times that from petrol engine exhaust. The concentration of smoke in the atmosphere can be estimated by drawing a known volume of air through a filter-paper and measuring the density of the resulting stain with a photo-electric reflectometer. Usually about 2 cubic metres of air are sampled per day. A calibration curve relating the density of the filter stain to the weight of smoke particles deposited on the filter-paper has been established for 'standard urban smoke'. Thus the concentration of smoke per unit volume of air can be calculated in terms of the 'standard smoke' equivalent.

** Total suspended particulare matter (TSP) is collected by means of a high-volume sampler. The sampler consists of a motor and blower enclosed in a shelter. The filter surface is arranged horizontally, facing upwards, and is protected by a roof that keeps out rain and snow and generally prevents the collection of particles larger than about 100 mm. Filters are made of glass or synthetic organin fibre. The air flow rates range from 1.1 to 1.7 cubic metres per minute. The amount of suspended particulate matter is calculated by dividing the net weight of the particulate by the total air volume sampled.

matter, there has been a strong association between these two pollutants. The particulars may at as carriers of SO_2 by delivering the SO_2 to the more sensitive areas (alveoli) of the lung.

Suspended particulate matter and sulphur dioxide are often regarded as the 'traditional' pollutants of urban areas. The highest levels of these pollutants occurred during the sulphurous smogs to which most large industrial cities have been subjected in the past. Although the worst smogs have now passed, some cities still experiences less intense smogs. The term 'smog' refers to a synthesis of smoke and fog. Smogs are caused by vast quantities of the pollutants being emitted from industry and from domestic sources (for example, coal fires, apartment incinerators) during periods when meteorological conditions fail to disperse the pollution away from the city. When winter anticyclonic conditions prevail-being characterized by calm or light winds, below-freezing temperatures and a restricted mixing depth due to a stable or inversion atmospheric lapse rate-little dispersal or dilution of pollutants occurs, and pollutions concentration may build up to high levels. Where an urban areas lies within a basin or valley, the low-level intense inversion of winter anticyclones forms a 'lid' and the hills surrounding the city form of the sides of a box into which massive quantities of pollutants are emitted. With the moisture added to the atmosphere by combustion process, the availability of vast quantities of condensation nuclei in the form of suspended particulars, and the low temperature which increase the relative humidity, fog formation is encouraged. The fog droplets readily dissolve sulphur dioxide to produce sulphurous acid, thereby adding to the potentially harmful nature of the smog. A polluted fog is less readily evaporated by solar radiation than a 'clean' fog, so the duration of the smog or pollution episode may be prolonged. The smog, has unhealthy and even dangerous high concentrations and may last several days until the anticyclone weakens or moves away from the region.

In urban areas the oxidation of SO_2 is a major source of particulate sulphates. Airborne acidic products of sulphur dioxide include sulphurous and sulphuric acid. These sulphur-containing aids in the atmosphere account for more than 60% of the acid found in low pH rain that acidifies lakes and rivers and also causes damage to buildings, vegetation, and wildlife.

Health Effects

The notorious sulphurous smogs which occurred in London in 1952 and 1962 and in New York in 1953, 1963, and 1966 clearly demonstrated that abrupt and substantial increase in the concentrations of suspended particulate matter and sulphur dioxide are positively associated with excess mortality. The individuals most susceptible were the elderly, the young, and individuals with chronic obstructive pulmonary disease and/or heart disease: for example, during or shortly after the four-day London smog of December 1952, an extra 4700 deaths occurred over and above the expected value. The increase in deaths from bronchitis was the largest single contributor to the rise in death rate, and deaths from other disease involving impairment of respiratory functions also increased. There was an increase in the number of deaths from heart disease, which could have been due to the additional strain placed on the heart by impairment of respiratory functions or to a direct effect. Peak daily concentrations of pollutants during the episode reached 6000 $\mu g/m^3$ of smoke together with nearly 4000 $\mu g/m^3$ of sulphur dioxide. Currently, with a few exceptions of suspended particulate matter or sulphur dioxide. Pollution control measure, together with socio-economic technological and energy-source changes, have dramatically reduced the levels of these pollutants.

The ability to smell SO_2 is highly variable, but there are reports that concentrations as low as 0.5 ppm can be detected. Acute inhalation of <1.0 ppm produced mild respiratory symptoms and small changes in the tracheobronchial region, including bronchoconstriction and altered air flow in both normal and asthmatic subjects. Changes in respiratory function in mild asthmatics has been reported at concentrations as low as 0.5 ppm SO_2 after only a 10 min exposure. These effects are transient and reversible and usually return to normal within 1 h. A variety of factors can modify this response such as dry or cold air, oral breathing, and exercise. Such effects have been duplicated in animals exposed to 2.0 ppm SO_2. An enhancement of the effects of SO_2 has been reported with concomitant exposure to particulates such as sodium chloride or ferrous sulfate.

In addition to change in lung function, SO_2 also alters normal respiratory defense system (Mucociliary clearance and nasal mucus flow) in both normal and asthmatic subjects at concentrations as low as 1.0 ppm. It is known that in diseases characterized by retarded clearance from the tracheal epithelium, e.g. chronic bronchitis, there is a predisposition to respiratory infections.

There is no evidence that long term continuous exposure to relatively high concentrations (5.0 ppm) of SO_2 produces any significant pathological changes in the lungs of exposed animals; however, at twice that concentration certain histopathological lesions become evident.

Effects on Vegetation

Vegetation may be adversely affected by excessive quantities of airborne particles. Particles cover leaves and plug stomata, thereby both reducing the absorption of carbon dioxide from the atmosphere and the intensity of sunlight reaching the interior of the leaf, and suppressing growth of some plants. Specific particles such as fluorides cause additional damage.

Acute injury to plants from sulphur dioxide initially takes the form of bleached patches on broad-leaved plants or bleached necrotic streaking on either side of the mid vein of parallel veined leaves. Long-term or chronic injury appears as a bleaching of the chlorophyll to give a mild chlorosis or discolouration (Yellowing) of the leaf in many plants. In other plants, the bleaching of the chlorophyll reveals the presence of red, brown or black pigments which are normally concealed. Whatever the form of the damage, the result is a reduction in growth and yield. Common plants susceptible to sulphur dioxide pollution include alfalfa, barley, cotton, lettuce, lucerne, rhubarb, spinach and sweet pea. Sulphur dioxide pollution does not always cause damage to vegetation because in sulphate-deficient areas, exposure to low levels of sulphur dioxide may be beneficial to plants by providing the missing sulphur. However, the sulphur dioxide may at the same time reduce the soil pH value, so requiring additional liming. Adverse effects of sulphur dioxide also occur via the effects of acid rain.

One plant particularly susceptible to ambient concentration of sulphur dioxide is the *lichen* and it is frequently used a bioindicator of sulphur dioxide levels. Sulphur dioxide interference with the photosynthesis in algal cells, eventually destroying the algae's chlorophyll; lichens are especially vulnerable perhaps because they contain relatively little chlorophyll. Experiments involving sulphur dioxide in solution show that different species are affected to different degrees by the same pollutant levels. Sulphur dioxide levels in the atmosphere can be determined by mapping the distribution, percentage cover and size of various species on particular substrates in an area, or from the fate of experimentally-transplanted lichens, or from differences in electrical conductivity between plants, since damaged plants develop weakened cell membranes and lose needed electrolytes.

Effects on Materials

Particles soiling fabrics, painted surfaces and buildings add to cleaning and replacement costs by reducing the life of materials and finishes. Particles may cause corrosion either by their intrinsic corrosiveness or by the action of absorbed corrosive chemicals, especially in a moist atmosphere. Soiling of building in cities in one of the more obvious manifestations of atmospheric pollution. Particles emitted from diesel exhausts are particularly effective in soiling as they have a high optical absorptance (blackness) and an oily nature.

Whereas suspended particulate matter soils or blackens the surface of materials, sulphur dioxide can produce substantial damage to materials. Limestone, sandstone, roofing slate and mortar of buildings and monuments can be severally damaged. The calcium carbonate in limestone and other building materials in readily converted into soluble calcium sulphate (gypsum). The increased volume associated with this chemical change causes sealing blistering and disintegration of the surface, with the loose material being washed away by rain. The most striking examples of this erosion are the historical monuments of Greece (for example, the Acropolis) and Italy (for example, the Coliseum and the Arch of Titus in Rome). They have withstood the influence of the atmosphere for hundreds or even thousands of years without any great changes, yet in the past few decades alone they have suffered very serious damage.

Fabric, leather, paper, electrical contacts, paint and medieval stained glass are all adversely affected by sulphur dioxide. Human-made textiles such as nylon are especially susceptible to sulphur dioxide or sulphuric acid aerosols. In addition, corrosion of metals, especially iron, steel, zinc, copper and nickel, is accelerated by the presence of sulphur dioxide which encourages the formation of sulphuric acid on the metal surface under moist conditions. Corrosion rates of metals in urban areas may be many times that experienced in rural environments. The lifetime of metals such as galvanized iron and steel can be extremely short in polluted areas with the need for costly replacement or frequent painting of important structures such as electricity pylons and bridges. Thus, for example, a galvanized steel plate with a coating 25 μm thick may require painting after three years in an industrial atmosphere while in a rural atmosphere this may last for 15–20 years before requiring painting.

Atmospheric Effects

One of the most obvious effects of pollutants in the atmosphere is the reduction in visibility caused by the absorption and scattering by solid and liquid aerosols. Relative humidity plays an important role because condensation occurs on hygroscopic particles at humidities as low as 70 per cent: the effect of suspended particulate matter alone (dry haze) is restricted to relative humidities below 70 per cent. During winter, under area encourages condensation leading to mist and fog formation. Fogs (defined as periods when visibility is less than 1000 meters) form more quickly and persist for longer in a polluted urban atmosphere. However, although fogs are in general more frequent in urban areas, dense fogs-says, with visibilities less than 400 meters-may be less frequent in city centers than in the suburbs. This occurs because fogs with visibility below 400 meters tend to have a higher moisture content than less dense fogs so do not form as readily in the slightly drier and warmer city centre.

The transport of sulphates and nitrates over long distances results in regional haze reducing visibility in areas distant from the emission sources of these secondary pollutants. Points to sulphates being the most important single contributor to visibility deterioration. The effects of secondary aerosols, especially sulphates, on visibility in the Southwest of the United States were strikingly demonstrated during the nine-month shut down of copper smelts in 1967-8. The copper smelts accounted for more than 90 per cent of the emissions of oxides of sulphur in the Southwest.

Sulphate and other submicron particles such as hydrocarbons are being increasingly transported further field to formerly pristine regions of the world such a the Arctic. Scientists are concerned not only about the effects on the fragile ecosystem but are also worried about the climatological effects of pollution. Preliminary research points to the Artic haze absorbing significant amounts of solar energy during the early spring months. It is possible that this absorption could cause a heating of the lower Artic atmosphere, causing consequent changes in the global climate.

There is no doubt that global levels of suspended particulates have increased since the turn of the century, but whether this will lead to a global warming or cooling disputed. Particles over low albedo areas such as ocean may increase the region's albedo and thereby prevent some of the solar energy from being absorbed by the oceans.

The major sources of lead in the environment which are of significance for the health of the communities arise from the industrial and other technological uses of lead. The major dispersive non-recoverable use of lead is in the manufacture of alkyl lead fuel additives. Other important sources of environmental lead relate to the manufacture of batteries, sheet and pipe, cable sheathing, older, shot and paint.

Alkellead compounds (tetramethyllead and teramethyllead) have been used as anti-known additives in petrol for over 50 years. Their use increased steadily into the 1970's after which the introduction of regulations on the maximum permissible concentration of lead in petrol together with the world oil and energy crisis, led to a decline in the amount of lead used. Lead compounds are added to fuel to prevent knocking, which occurs when there is spontaneous ignition of the petrol and air mixture in the corner of the cylinder furthest away from the spark plug (the end-gas zone). For a given design of engine, the presence or absence of knocking is determined by the combustion quality or 'octane number' (percentage of a very high knock-resistant type of hydrocarbon) of the petrol used the higher the octane number, the higher the compression ratio at which the engine can be run without knocking. The addition of organic lead compounds (lead alkyls) is a cheaper and more convenient method of boosting the octane quality of petrol than by intensive refining. The presence of lead in the fuel has the effect of delaying the abnormal state of oxidation in which the end-gas ignites, and hence allows engines to operate at higher compression ratios and under greater loads before knocking occurs. If tetraethyllead and teramethyllead are not employed, the choice to the motor industry is either to accept increasing refining costs of the petroleum (to increase the proportion of knock-resistance hydrocarbons) or to make use of reduced octane quality in vehicles.

The concentration of lead in air may vary from 2 to 3 $\mu g/m^3$ in large cities with dense motor traffic, to less than 0.2 $\mu g/m^3$ in most suburban areas and still less in rural areas. Concentrations are highest along highways during rush hours. Excessive concentrations may also be measured in the vicinity of industrial sources such as lead smelters: Many nations have adopted an annual mean lead concen tration 2 $\mu g/m^3$ as a long-term health-related goal for air-pollution control purposes.

People are exposed to varying levels of lead in the atmosphere according to where they live and work and how they commute. However, the to where they live and work and how they commute. However, the contribution of airborne lead by direct inhalation is but one pathway by which airborne lead contributes to the total lead intake in people (Fig. 3.2). In addition to direct or indirect airborne lead contributions, there several other ways in which lead levels in the body are increased. In areas where the water is sold (low in calcium and magnesium) and where, at the same time, lead pipes and lead-lined water storage tanks are used, a percentage rise in the body burden of lead occurs, which is reflected in elevated values of lead in the blood. Since lead in water ingested independently of food is more readily absorbed, it may provide a relatively greater contribution to the blood lead level than lead in food. Lead concentration in food is highly variable, with foods and wines which are stored in lead soldered cans or lead-glazed pottery being especially high in lead content. For infants and young children, an additional source of lead may come from miscellaneous lead-containing objects which are licked, chewed or eaten, as well as from hand-to-mouth contact with lead-contaminated dust.

Health Effects

Airborne lead can have a health impact not only through inhalation but also through the ingestion of particulate lead which can be deposited on food or fur of organisms and be taken in orally during eating or grooming. The respiratory absorption of lead is

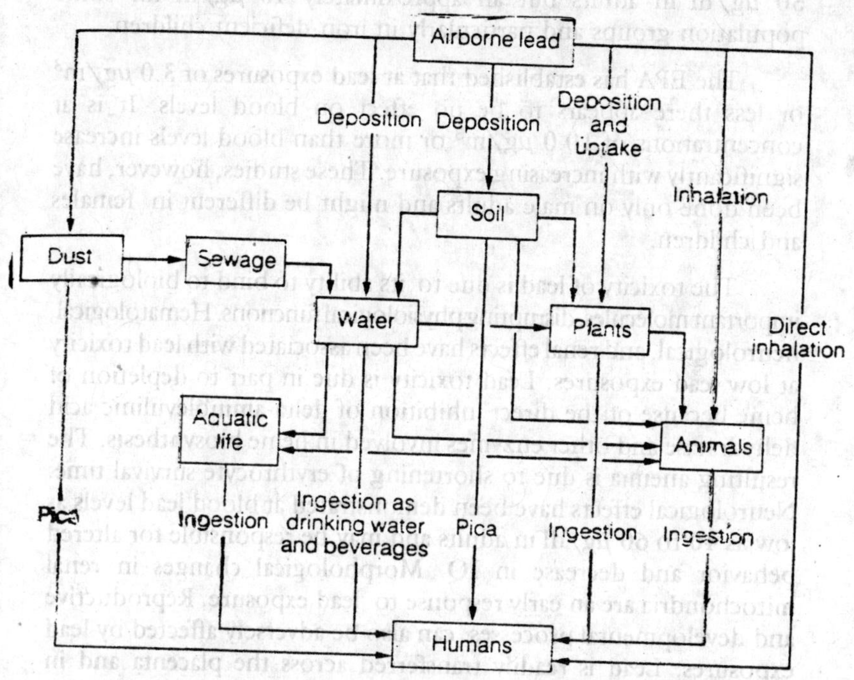

Fig. 3.2 : Contribution of airborne lead to total lead intake

dependent on two processes: the deposition of particulates of lead into the lung and the absorption of lead into the bloodstream. Deposition is dependent on particulate size, ventilation rate, and respiratory morphology. Approximately 30 to 50% of the lead inhaled by an adult will be deposited deep into the respiratory tract. Systemic absorption from the alveoli of the lung occurs directly, while lead deposited in upper respiratory tract can be absorbed by swallowing and absorption from the gut. Administration of pB tracers to human volunteers indicates that nearly all the lead deposited in the lower respiratory region is absorbed. Absorption of lead varies significantly with the age, sex, and nutritional status of the organism. Lead uptake is much greater in children than adults which is thought to reflect a higher deposition rate.

Clinical signs of lead poisoning are fairly well established. Anemia is a characteric early toxic effect in people, with a slight decrease in the haemoglobin levels occurring at a mean level of about

50 $\mu g/dl$ in adults but an approximately 40 $\mu g/dl$ for some population groups and particularly in iron-deficient children.

The EPA has established that at lead exposures of 3.0 $\mu g/m^3$ or less there appears to be no effect on blood levels. It is at concentrations of 10.0 $\mu g/m^3$ or more than blood levels increase significantly with increasing exposure. These studies, however, have been done only on male adults and might be different in females and children.

The toxicity of lead is due to its ability to bind to biologically important molecules, disrupting physiological functions. Hematological, neurological, and renal effects have been associated with lead toxicity at low lead exposures. Lead toxicity is due in part to depletion of heme because of the direct inhibition of delta-aminolevulinic acid dehydratase and other enzymes involved in heme biosynthesis. The resulting anemia is due to shortening of erythrocyte survival time. Neurological effects have been demonstrated at blood lead levels as low as 40 to 60 $\mu g/dl$ in adults and may be responsible for altered behavior and decrease in IQ. Morphological changes in renal mitochondria are an early response to lead exposure. Reproductive and developmental processes can also be adversely affected by lead exposures. Lead is readily transferred across the placenta and in rodent studies fetotoxicity has occurred at a concentration of 10 $\mu g/m^3$ (U.S. PEA, 1986). Epidemiologic studies indicate that fetal exposure to lead may have undesirable effects on mental development of the newborn and length of the gestation period. In the male, lead may have an adverse effects on the development of sperm and the seminal vesicles.

Lead has been classified by EPA as a probable human carcinogen although this is not fully established. However, there is no epidemiological evidence that workers exposed to lead have a higher incidence of renal cancer than the normal population.

4

Petroleum and Solvents

PETROLEUM

Composition of Petroleum

The natural composition of petroleum is very complex and, varies between petroleum reservoirs. In general, petroleum is a complex mixture of hydrocarbons i.e., molecules made up of only carbon and hydrogen and non-hydrocarbons molecules made up of carbon, hydrogen, and other elements such as sulphur, nitrogen, and oxygen that forms from the partial decomposition of biogenic materials. The slow break-down process, known as diagenesis, produces a range of hydrocarbons and hydrocarbon complexes significantly altered from the structure found in the original biomass. The resulting breakdown products occur in a combination of three phases.

1. Gaseous form—natural gas
2. Liquid Form—crude oil
3. Solid form-—tar or bitumen

Each petroleum reserve is a unique combination of biomass breakdown products. Hence petroleum each have a unique compositional complexity, with variations occurring within the individual petroleum reservoir.

Since there are compositional differences in petroleum no specific definition or composition statement is valid for all crude oils. A broad functional definition of petroleum hydrocarbons is that hydrocarbons are primarily composed of many organic compounds of natural origin and low water solubility. The molecular compositional differences in crude oils result in physical and chemical properties that are unique for each type of oil. These differences are important for refinery processing and for predicting the potential impacts to the environment from oil spills.

Crude oils are used for the production of fuels and lubricants for transportation and energy applications and in the petrochemical industries. The refinery processes that convert crude oils to useful products include 1) distillation techniques that separate petroleum products with different boiling point (bp) ranges and 2) catalytic cracking processes that create usable products from the unusable components of oil. The general range for distillation products, are

Boiling Point Range	Petroleum Distillation Ranges
<20°C	Natural gas
20–200°C	Straight-run gasoline
185–345°C	Middle distillates—kerosene, jet fuel, diesels, heating oil
345–540°C	Light to heavy lubricating oils and waxes, feed stock for catalytic cracking to gasoline
>540°C	Residual oil

Elemental Composition. Most of the chemical components in petroleum are made up of five main elements. The most abundant element by weight is carbon. Most oils contain a low percentage of sulphur. Oil with a higher percentage of sulphur, are associated with a carbonate rock source that lacks inorganic iron to bind with the sulphur. The nest most common elements in oils are nitrogen and oxygen.

A wide range of metals are found in trace amounts in crude oils. The elemental and trace metal components of petroleum are listed in Table 4.1 and 4.2.

Molecular Composition. The following classification scheme breaks petroleum components into five major groups. The subcat

Table 4.1 : Elemental Composition of Crude Oil

Element	% Weight
Carbon	82 – 87
Hydrogen	11 – 15
Sulfur	0 – 8
Nitrogen	0 – 1
Oxygen	0.0 – 0.5

Table 4.2 : Trace Metal Composition of Crude Oil

Element	Concentration (ppm)
Aluminium	11
Antimony	11
Barium	3
Boron	<1
Cadmium	<1
Calcium	37
Chromium	3
Lead	2
Magnesium	8
Nicket	12
Phosphorus	98
Silicon	7
Tin	13
Vanadium	38
Zinc	7

egories are based on the majority of the compounds in that subcategory.

Saturate hydrocarbons	Normal paraffins (straight-chain hydrocarbons) Isoparaffins (branched-chain hydrocarbons) Cycloparaffins or naphthenes (cyclic saturated and partially saturated hydrocarbons)
Aromatics	Aromatic hydrocarbons (AH) Polynuclear aromatic hydrocarbons (PAH) and their C_1 to C_4 alkyl homologs
Polar Compounds	Sulfur-containing aromatic compounds Nitrogen-containing aromatic compounds

	Oxygen-containing aromatic compounds
Porphyrins	**Complex large cyclic carbon structures derived from chlorophyll and characterized by the ability to contain a central metal atom (trace metals are commonly found within these compounds)**
Asphaltenes and resins	**Composition is dependent upon source (these structures have the highest individual molecular weight of all crude oil components and are basically colloidal aggregates)**

The saturated carbon bond, (that is carbon bound to four other atoms) in hydrocarbon molecules can be arranged as either straight-chained, branched, or cyclic hydrocarbon structures (Fig.4.1). The straight-chain compounds are found in a homologous series (same molecular structure separated by only CH_2 groups) in most crude oils, with carbon numbers ranging from 5 to 6 up to 30 or more. These compounds are generally the major components of

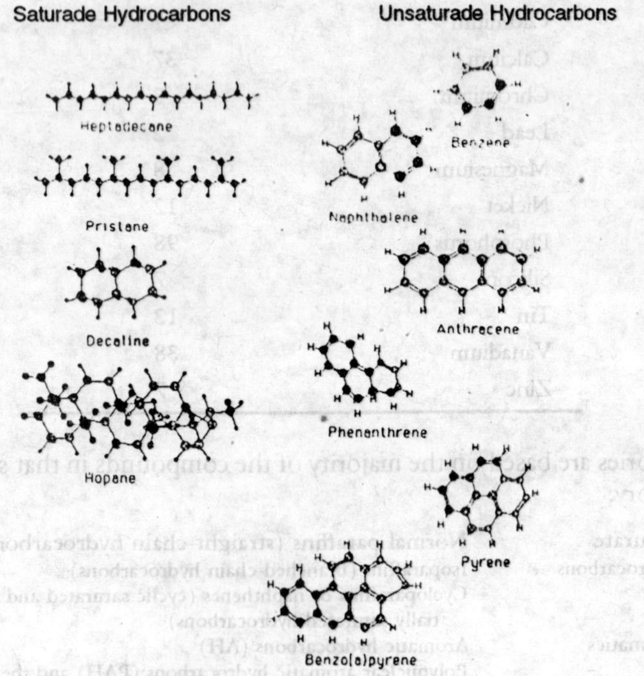

Fig. 4.1 : Chemical Structures of Hydrocarbons

crude oils and are known as normal paraffins. The branched chain compounds, or isoparaffins, are primarily methylated straight chain hydrocarbon. From nine carbons and higher, most isoparaffins are isoprenoid originating from a side chain of the chlorophyll molecule, with a methyl group attached to every third carbon in the chain. These branched and isoprenoid hydrocarbons are generally degraded more slowly than the normal paraffins by microorganisms. Two common paraffins found in most petroleums are heptadecane, a straight 17-carbon chain molecular structure, and pristane, a branched 19-carbon chain isoprenoid structure. These two components have molecular weights of 240 and 268, respectively, although their bp and gas chromatographic retention times are similar. Their microbial degradation rates in the environment are very different. Heptadecane rapidly degrades in the environment, while pristane remains and is often used as a biomarker for weathering and degradation studies. Other abundant structures in most fresh crude oils are the cyclic hydrocarbons. The two basic cyclic structures are cyclohexane and cyclopentane, with five and six carbon rings, respectively. Though combinations of these two-rings structures assembled with up to five rings, many complex molecular are formed during petroleum diagenesis. Two of the larger structures that are resistant to weathering processes are the triterpanes and the styrenes, with a subclassification of happens. These compounds, which are not degraded by microorganisms to any appreciable extent, are also used as degradation biomarkers.

Carbon atoms that are bound to three or fewer other atoms in hydrocarbon molecule are referred to as "unsaturated". Their physical structure is a straight, branched, or aromatic form (Fig. 4.1). Crude oils do not contain unsaturated hydrocarbons other than the aromatic structures. However, refined petroleum products may contain unsaturated compounds with one or more double bonds in their molecular structure. As with the saturate petroleum hydrocarbons, aromatic ring structure in crude oils range from one-to five-ring combinations. Two or more five-or six-member carbon rings are fused together to form polycyclic aromatic compounds (PACs). These petroleum aromatic hydrocarbons have alkyl groups substitution on their ring structures. The alkyl groups generally have from one to four saturated carbon atoms and thus can produce many different structural isomers and homologs for each aromatic hydrocarbon family.

The toxicological implications from petroleum occur primarily from exposure to or biological metabolism of these aromatic structures. The most acutely toxic compounds are the single aromatic ring compounds including benzene, toluene, and the xylenes. All are volatile compounds with relatively high water solubility for insoluble compounds. Therefore, both atmospheric and hydrospheric impacts must be assessed when considering toxic implications from a petroleum release containing significant quantities of these single-ring aromatic compounds. The dibenzene ring structures, or naphthalene and its homologous series, are less acutely toxic than benzene but are more prevalent for a longer period during oil spills. Table 4.3 shows a comparison of Threshold Limit Value (TLV).

The polar compounds are also known as the nonhydrocarbons. For these compounds, the primary functional group is associated with nitrogen, sulphur, or oxygen atoms. The classification applies to sulphur compounds such as cyclic sulphides and thiophenes, basic and neutral nitrogen compounds, and oxygen compounds, primarily naphthenic acids.

The porphyrins, asphaltene, and resin compounds in petro leum are considered the residual oil. During weathering processes. This fraction is the last to degrade, and its persistence over years has been noted. This residual fraction is commonly used in roofing

Table 4.3 : Aromatic Compounds in Petroleum

Compound	Flash Point (°F)	TLV (ppm)	Solubility (mg/l at 25°C)
Benzene	12	10	1,800
Toluene	40	10	470
Xylenes	63	100	198
Naphthalene	174	10	31.7
Phenol			82,000
Benzaldehyde			3,300
Benzoic acid			2,900
Methylbenzene			515
Biphenyl			7.5
Benzo(a) pyrene			<0.01*

* Almost insoluble

compounds, hot melt adhesives, and heating oils.

Characteristics of Petroleum

The physical characteristics are determined by the characteristics of the individual components within the petroleum and their relative quantities. These physical characteristics, include density, specific gravity, viscosity, pour point, and flash point. They are generally derived from the total composition of petroleum and petroleum by-products as a weighted average of the properties of individual compounds in the complex mixture.

Density. Density is the mass per unit volume of a substance. The units expressed are grams per cubic centimeter for solids, g/cm^3 and liquids and grams per liter g/L for gases. This unit is based on the definition that 1 cm^3 water weights 1 g. g/L.

Specific Gravity. Specific gravity relates the density of a substance to the density of a standard substance. Water is commonly used as the standard for solids and liquids and has specific gravity of 1.0000. For solids and liquids, the specific gravity term is generally equivalent to density (assuming the density of water is 1.0000). In the petroleum industry, specific gravity is generally expressed in "degree API" in accordance with a scale established by the American Petroleum Institute (API). The relationship between specific gravity and degree API is defined as

Specific gravity = 141.5/(131.5 + degrees API)

Viscosity. Viscosity is determined by internal friction and is measured as resistance to flow and in defined as the force required to move a 1 cm^2 planar surface area above another planar surface at a rate of 1 cm s^{-1} when the two surfaces are separated by a layer of fluid 1 cm in thickness. The unit of viscosity is the poise. The ratio of absolute viscosity to density is called the kinematic viscosity and is measured in stocks (St) or centistokes (cSt).

Pour Point. The lowest temperature at which an oil can be poured is the pour point. Pour points for crude oils generally range from to 43°C. If the pour point of the oil is higher than the temperature of the environment, the oil will tend to aggregate rather than spread as a liquid.

Flash Point. The flash point is the temperature at which a liquid or volatile solid gives off vapors sufficient to form an ignitable mixture with air near the surface of the liquid or solid. The flash point temperature is determined by the composition of the more volatile components in petroleum. Petroleums that primarily contain volatile compounds of low molecular weight have low flash points while petroleums containing primarily compounds of high molecular weight and low volatility have high flash points.

These physical characteristics provide insight into the collective molecular composition in crude oils. For example, a "light" crude oil is one with high API specific gravity and low viscosity and pour point temperature when compared to other oils. This generally implies a higher percentage of saturates and lower asphaltene and polar content in the petroleum mixture. Asphaltenes are substances in petroleum that are insoluble in solvents of low molecular weight such as pentane or hexane. These compounds are composed of very large cyclic and planar molecules and are solids at normal temperatures. Consequently, oils that have high asphaltene contents are very viscous, have a high pour point, and are generally nonvolatile in nature.

Solubility in Water. One of the most important properties in determining the environmental fate of pollutants is their solubility in water. Solubility of a substance also determines the routes of exposure that are possible. Certain functional groups enhance the solubility of benzene in water. For instance –OH in phenol; –CHO in benzaldehyde and –COOH in benzoic acid (Table 4.3).

Vapour Pressure. Vapour pressure is the pressure exerted by a vapour in equilibrium with the liquid from which it is derived at a given temperature (usually 20°C). This value is significant for predicting a volatility of crude oils and refined products. Refined light petroleum products, under increased temperature and pressure, from explosive mixtures in air.

Weathering. "Weathering" includes the physical and chemical changes that crude oils and refined petroleum products undergo as a result of their interaction with the environment. Of the possible environments impacts, the hydrosphere presents the most complicated and interactive zone.

The fate of oil in an environment is governed by the properties of the individual constituents in the oil. The factors affecting the fate of oil spilled or released in an aquatic environment include spreading, evaporation, dissolution, dispersion, emulsification, absorption by sediments, and photochemical and microbial degradation.

The ultimate fate of petroleum spilled in the marine environment is degradation primarily biological and, to a lesser extent, photolytic gradation. The rate at which these processes occur is controlled by the following factors, 1) the physical and chemical composition of the oil, 2) the exposure to physical processes, 3) abiotic environmental factors, 4) and the preexistence or absence of hydrocarbon-degrading microorganisms (bacteria, mold, yeast, and fungi).

Most chemical breakdown processes affecting whole oil occur at the oil/water interface and in the atmosphere. Chemical actions in aquatic system include biodegradation and photolytic degradation, which are surface-active processes. Therefore, the amount of surface area affects the rate of degradation. The factors that hinder degradation are 1) emulsification, which slows biodegradation because the stable "mousse" formed has a smaller surface area compared to very thin oil sheens, 2) cool weather, often results in slowed rates of microbial activity, which results in less consumption or degradation of the oil.

The microbes utilize the oils as a carbon source which is converted to energy, biomass, carbon dioxide, and water. Physical processes such as high energy storms increase the dispersion of oil, thus increasing surface areas and generally reducing the persistence of oil in the environment.

Weathered oil can be distinguished from fresh or unweathered oil by the loss of the low molecular weight constituents such as the normal hydrocarbons less than n-C_{12} (a 12-carbon molecule), the alkylbenzenes, and the naphthalene homologs. These initial changes are primarily due to evaporative losses and, to a lesser degree, dissolution. For example, less than 5% of benzene is generally lost to dissolution, while more than 965% is lost to evaporation. The constituents of low molecular weight are more volatile and more water soluble than the constituents with high molecular weight. Many refined products such as bunker oil appears as "weathered" crude oils, since they lack the compounds of lower molecular weight. The high viscosities of these refined products may make them very resistive to physical weathering in much the same way as weathered crude oils and may also make them slow to further degrade.

Toxic Effects of Petroleum

The toxic effects of crude oils and refined products are difficult to assess even though the toxicity of each individual component is known. The problems arise due to the following difficulties:

1. there is a lack of information on additive, synergistic or antagonistic effects of various compounds of crude oils.
2. The chemical composition of each crude oil and petroleum product, which varies significantly, can have diverse effects on organism within the same ecosystem.

Benzene and naphthalene components in fresh crude oil are considered to be the most toxic and most abundant compounds during the initial phases of petroleum spills. The naphthalene- and benzene-derived compounds, other than benzene, are considered noncarcinogenic. At the initial stages of the release, when these compounds are present at their highest concentrations, acute toxic effects are most common. These noncarcinogenic effects range from subtle changes in detoxifying enzymes to liver damage and interference with reproductive behavior. The severity of the effects depends on the organism being exposed, the concentration of the compounds, and the route of exposure.

As oil weathers, it becomes less acutely toxic, but it remains toxic via injection because it contains polynuclear aromatic hydrocarbons (PAHs) of high molecular weight. Many of these compounds, including at least one known human carcinogen, benzo(a)pyrene, have been identified as mutagenic; other are listed as experimental carcinogens. The remaining compounds may become enriched in the residual bulk oil due to the loss of the volatile, soluble, and easily biodegraded compounds. It is important to point out that crude oils contain primarily the alkyl homolog of aromatic compounds and relatively small quantities of the unsubstituted "parent" aromatic ring structures. It is these unsubstituted aromatic structures that are potentially responsible for the majority of the known toxic impacts of crude oils.

Routes of Exposure. The possible routes of exposure for the various classes of compounds found in petroleum depend upon their physical and chemical characteristics. If a spill or release occurs on land, containment practices are readily available; if the spill or release occurs on a body of water, containment and cleanup operations are affected by the availability of regional resources and by the weather. Petroleum is naturally weathered according to its physical and chemical properties, but during this process living species within the local environment may be affected via one or more routes of exposure, including ingestion, inhalation, skin contact, and, to a much lesser extent, bioconcentration through the food chain. The more water-soluble petroleum compounds are dispersed in the water column, where they may be ingested or inhaled by marine species. Exposure of marine and avian populations to petroleum remaining on the water surfaces or shorelines occurs by ingestion through preening and by skin contact. The bioconcentration process applies to the predator/prey relationship, primarily by inspections of oil-soiled animals.

Experimental biological studies and studies of past petroleum spills indicate that physical contact with weathered oil may reduce the natural protective barrier of birds and marine mammals to the elements resulting in hypothermia.

Bioavailability, sensitivity, and bioaccumulation affect the toxic responses of a given species. Sensitivity to petroleum varies for each species, and differing degrees of sensitivity occur at different stages of life.

Birds. Bird species with water habitats are the species most commonly affected by oil spills and releases. These birds are dependent upon the marine or freshwater environment for food sources and protection and are most likely to be exposed to petroleum. The ingestion of petroleum-contaminated food sources is not common for most of these birds, though scavenging or hunting birds may feed on petroleum-impacted animals washed ashore. Exposure of water fowl occurs primarily through the fouling of plumage. Oil disrupts the fine strand structure of the feathers, resulting in loss of water repellency and in decreased body insulation. As the oiled plumage becomes matted, water penetrates the feathers and chills the body; the combined results are loss of buoyancy and possible hypothermia.

Petroleum on the external surface of eggs has also been noted to have toxic effects. Coatings of some crude and fuel oils covering less than 10% of the egg surface have been shown to cause reduction in hatching, teratogenicity, and stunted growth.

Marine Mammals. Marine mammals and pelagic species have similar routes of exposure: inhalation, surficial contact, and ingestion. The marine species include cetaceans, pinnipeds, and sea otters. Very rarely are the larger mammals exposed to petroleum slicks on the water; they have been known to alter their routes to avoid petroleum. It is the sea otter that resides in a localized region that is most affected and has the most reported deaths from direct and indirect exposure.

Inhalation of oily water more frequently for those species that feed on the water surface or in shallow impacted waters. This leads to absorption by the circulations systems, irritation, and possibly permanent damage to respiratory surfaces and mucous membranes. The sea otters are the most sensitive, with effect to the lungs and nervous system.

Surficial contact has little or no effect for most whales due to the insulating qualities of their blubber and skin. The adult seals, sea lions, and walruses have insulating qualities similar to the whales and are not greatly affected by surficial contact. The only surficial membranes noticeably affected are the mucous eye membranes. The hypothermic response, similar to that of water fowls, occurs when petroleum-matted fur has reduced heat insulation. Affected animals must expend much more energy to maintain a normal body temperature, stressing the body systems. The majority of the sea otter deaths during petroleum spills have been associated with this hypothermic response.

Ingestion occurs through direct routes, by consumption of contaminated food, or indirectly, by grooming in attempts to remove the petroleum matting. The results are both acute and chronic and include kidney failure, destruction of intestinal lining, neural disorders, and bioaccumulation. No biomagnification instances have been reported, presumably due to the effective assimilation of petroleum compounds from food sources.

SOLVENTS

Introduction

Solvent are chemically diverse group of compounds, that are usually liquids, a wide range of specific gravities, viscosities, vapor densities, and flash points. Solvents are not highly lipophilic, but lipophilicity is a toxicologically significant physical characteristic of some types of solvents. The distribution of the solvent of target organs in an organism is determined by cardiovascular efficiency and the physicochemical properties of the solvent. Partition coefficient. Blood pH should not affect the transfer of most solvents across cellular membranes since solvents are usually not ionized in solution. The extent and rate of transfer will thus depend mostly upon their partition coefficient. Absorption of solvents from the lungs is unobstructed, and dissolution in the blood occurs in accordance with their solubility in blood expressed as an Ostwald coefficient.

Solvents generally have high vapor pressure which result in the formation of vapours; therefore, the primary exposure route of solvents is by inhalation. The intake rate of solvents received by inhalation depends on the rate and depth of respiration. The toxicity of solvent vapours is therefore expressed as a LC_{50}, air concentration resulting in the depth of 50% of exposed animals, or the concentration in air multiplied by the time of exposure which results in death in 50% of the animals.

Exposure through the skin is frequent. As solvents are lipophilic substances, it is expected that they are absorbed through the skin into the bloodstream. Systemic toxicity can occur from absorption of some solvents. (absorption may occur by absorption and inhalation simultaneously)

Although the physical and chemical properties of solvent differ most solvents exhibit a depressing effect upon the central nervous system (CNS) and are therefore classified as CNS toxicants. CNS symptoms produced by solvents are similar to those of general anesthetic agents suggesting that the mechanism of action is that of a physical interaction of the solvent with the membranes of certain cells within the CNS that results in changes in cell membrane permeability to ions. It appears that the depressant effect of solvents are related to concentration. The concentration that is required to produce an effect is much greater than would be necessary were a receptor mechanism involved.

The effect of organic solvents on the CNS are related to the concentrations of the chemicals that reach the brain. Symptoms of CNS depression exposure to solvents are nausea, dizziness, euphoria, confusion, and loss of coordination at low concentrations and, convulsions, coma, and death at high concentrations. It is interesting that the metabolites of some solvents such as n-hexane and methanol are responsible for more severe toxicological effect that the parents chemical. In addition to other toxic effect, some solvents or their metabolites affect specific organs that are called target organs. Examples of solvents and target organs: Table 4.4 some solvents and their targets organs are given in Table 4.4.

Table 4.4 : Some Solvents & their Target Organisms

Solvent	*Target Organ*
Methanol	Ocular nerve or retina
Chlorinated hydrocarbons	Liver
Ethanol	Liver, brain
Hexane	Nerves
Ethylene glycol ethers	Reproduction
Benzene	Bone marrow
Carbon disulphide	Nerves

With the enormous use of chemicals in industry, many workers are exposed to solvents, either in the ambient air or by skin contact. Therefore, it has become necessary for the sake of health and safety to attempt to establish air concentration limits for handling procedures for chemicals, including solvents. Establishment of these limits is of benefit not only to the workers and management of health and economic reasons but also provides guidelines for the general population which may be exposed to hazardous waste sites, spills, etc. The establishing of threshold limit values (TLV) has, by necessity, involved toxicological studies and reevaluations of existing data about the toxic effects of chemical on person in situations of prolonged exposure.

TOXICOLOGY OF SOLVENTS

Alkanes (Saturated Aliphatic Hydrocarbons)

This group of solvents contains hydrocarbons which are sometimes referred to as paraffins or alkanes. They have the general formula C_nH_{2n+2}. The lower members are gases; the solvents are above C_5H_{10} and include pentanes to hexadecanes (C_5H_{10} to $C_{19}H_{28}$).

Petroleum and petroleum products are recognized as the source of many of the chemicals of this group. Petroleum products are used as motor fuel, lubricants, solvents, and in the synthesis of many different chemicals.

Most members of this groups are highly lipophilic and are distributed rapidly throughout the body. They are absorbed readily by the lungs and skin, and defat the skin upon contact. Members of this group are not excreted very readily, and they are metabolized to their more hydrophilic derivatives, alcohols.

Alkanes are depressants to the CNS. Their principal toxicity effect in acute toxic ingestion is that of aspiration of the hydrocarbon into the lungs either at the time of ingestion or upon regurgitation. The aspiration on only 0.1 ml into the lungs can result in chemical pneumonitis. The probability of a hydrocarbon solvent being aspirated is related to the viscosity of the solvent.

The symptoms of toxicity induced by alkanes include narcosis, dizziness, headache, nausea, confusion, and in very carbon number from C_1 to C_8 (octane) exhibit increasing toxic effects on the CNS. Octane is about 15 times more potent a narcotic than pentane (Table 4.5). Sudden death have been reported in persons who have inhaled

Table 4.5 : Toxicological Properties of Some Alkyl Hydrocarbons

Hydrocarbon	*Animal*	*Exposure*	*Toxic Effects*
Pentane	Mice	90,000–120,000 ppm (5–60 min)	Narcosis
Hexane	Mice	30,000 ppm	Narcosis
Hetane	Mice	10,000–15,000 ppm (30–60 min)	Narcosis
Octane	Mice	6,600–13,700 ppm (30–90 min)	Narcosis

vapours of hydrocarbons such as gasoline.

n-Hexane is widely in industry in the synthesis of products such as plastics, paint thinners, lacquers, and glue.

n-Hexane is metabolized in the liver by the mixed function oxidase system to 2-hexanol, which in turn is oxidized to 2.5-hexanediol which is oxidized to 2,5-hexanedione. The metabolism of hexane is shown in Fig 4.4.

n-Hexane is a CNS depressant. At low exposure levels, head-ache and anoxia have been reported. Confusion, stupor, and coma can result from exposure to high concentrations of *n*-hexane. Chronic exposure to *n*-hexane produce a polyneuropathy. Incidence of polyneuropathy has been reported in workers who were exposed to ambient air containing high concentration of n-hexane in glue vapours from recreational use. A hexane-induced polyneuropathy is characterized by fatigue, muscular weakness, and distal paresthesia (tingling sensation) in upper and lower extremities. The mechanism by which hexane and hexane metabolites actually damage neurons is not known. 2.5,-Hexanedione has been shown to interact with several cellular molecules which would result in alteration of the neuronal functions.

Aromatic Hydrocarbons

This group of solvent includes benzene and derivatives of benzene.

Benzene. Benzene is used as an additive for gasoline and other fuels and as a solvent by many industries, that is, paint, plastic, and rubber. Ninety percent of the benzene produced is used in the manufacture of medicinal chemicals, dyes, and other organic compounds.

Benzene has been classified as a priority toxic pollutant, a hazardous substance and a hazardous waste constituent.

The principle route of exposure to benzene is by inhalation. About 3- to 80% of the benzene inhaled in absorbed into the circulating blood. Benzene can be absorbed through the skin. Benzene is distributed widely in the body with the greatest concentration being found in the fat. About 10 to 15% of absorbed benzene is returned to the lungs and excreted unchanged by expiration.

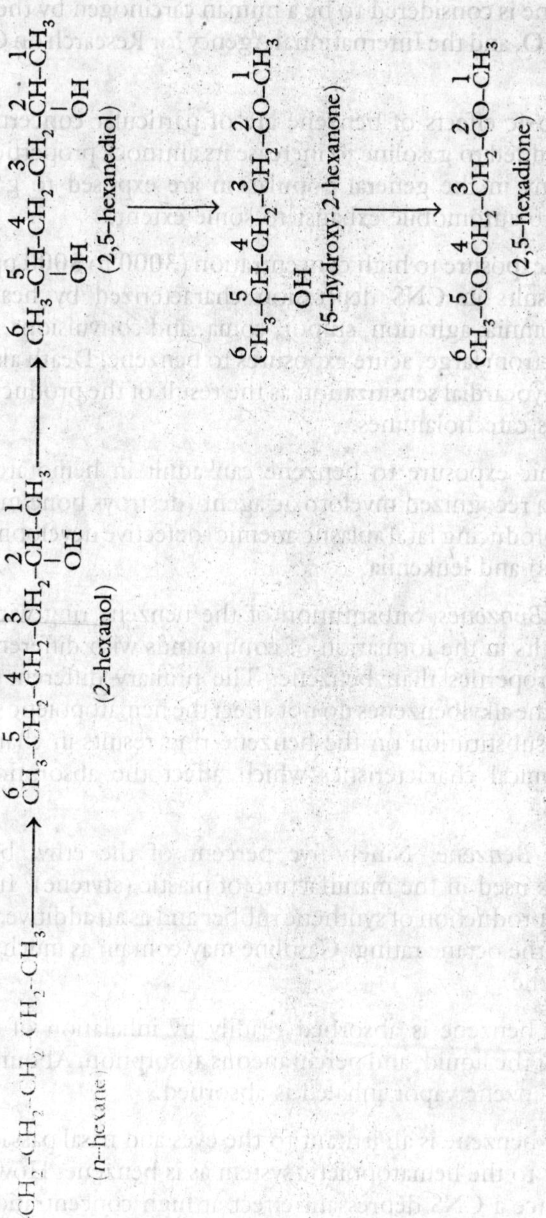

$$CH_3{-}CH_2{-}CH_2{-}CH_2{-}CH_2{-}CH_3 \longrightarrow \overset{6}{C}H_3{-}\overset{5}{C}H_2{-}\overset{4}{C}H_2{-}\overset{3}{C}H_2{-}\underset{\displaystyle OH}{\overset{2}{C}H}{-}\overset{1}{C}H_3 \longrightarrow \overset{6}{C}H_3{-}\underset{\displaystyle OH}{\overset{5}{C}H}{-}\overset{4}{C}H_2{-}\overset{3}{C}H_2{-}\underset{\displaystyle OH}{\overset{2}{C}H}{-}\overset{1}{C}H_3$$

(n-hexane) (2–hexanol) (2,5–hexanediol)

$$\downarrow$$

$$\overset{6}{C}H_3{-}\underset{\displaystyle OH}{\overset{5}{C}H}{-}\overset{4}{C}H_2{-}\overset{3}{C}H_2{-}\overset{2}{C}O{-}\overset{1}{C}H_3$$

(5–hydroxy-2-hexanone)

$$\downarrow$$

$$\overset{6}{C}H_3{-}\overset{5}{C}O{-}\overset{4}{C}H_2{-}\overset{3}{C}H_2{-}\overset{2}{C}O{-}\overset{1}{C}H_3$$

(2,5–hexadione)

Fig. 4.1 Metabolism of *n*-hexane

Benzene is considered to be a human carcinogen by the EPA, OSHA, WHO, and the International Agency for Research on Cancer (IARC).

The toxic effects of benzene are of particular concern since benzene is added to gasoline to increase its antinock properties, and because many in the general population are exposed to gasoline fumes and to automobile exhaust to some extent.

Acute exposure to high concentration (3000 to 5000 ppm) of benzene results in CNS depression characterized by headache, nausea, insomnia, agitation, stupor, coma, and convulsions. Death has resulted from large, acute exposures to benzene. Death also may be due to myocardial sensitization as the result of the production of endogenous catecholamines.

Chronic exposure to benzene can adult in hemotatoxicity. Benzene is a recognized myelotoxic agent (destroys bone marrow) capable of producing fatal aplastic anemic (defective development of erythrocytes) and leukemia.

Alkyl Benzenes. Substitution of the benzene ring with alkyl groups results in the formation of compounds with different toxicological properties than benzene. The primary different is that apparently the alkylbenzenes do not affect the hematopoietic system. Also, alkyl substitution on the benzene ring results in changes in physicochemical characteristics which affect the absorption and metabolism.

Ethyl Benzene. Ninety-five percent of the ethyl benzene produced is used in the manufacture of plastic (styrene). It is also used in the production of synthetic rubber and as an additive to fuels to increase the octane rating. Gasoline may contain as much as 20% ethyl benzene.

Ethyl benzene is absorbed readily by inhalation of vapors, ingestion of the liquid, and percutaneous absorption. About 64% of the ethyl benzene vapor inhaled is absorbed.

Ethyl benzene is an irritant to the eyes and nasal passages but is not toxic to the hematopoietic system as is benzene. However, it does produce a CNS depressant effect at high concentration.

Toluene. Toluene (methyl benzene), xylene (dimethyl benzene) and mestylene (trimethyl benzene) are most frequently used solvents. Their toxicology is of particular importance since they are produced used and disposed in large quantities.

Toluene is more lipid soluble and less volatile than benzene. Toluene is used in the chemical industry to produce several other chemicals such as solvents for paints, lacquers, coatings, and glues and fuels in automobiles and aircraft. With the widespread use of glues, thinners, etc., there is possibility of the exposure of a large number of the general population.

Toluene is rapidly absorbed by inhalation with peak blood concentrations occurring is 15 to 30 min after inhalation. Toluene distributes throughout the body and accumulates in the adipose tissue.

Toluene (80%) is metabolized in the benzyl alcohol and then to benzoic acid which is conjugated to produce hippuric acid. Some toluene is excreted unchanged by expiration from the lungs. Toluene has a termination half-life of 15 to 20 h.

Symptom of acute exposure to toluene include headache, dizziness, fatigue, muscular weakness, collapse, and coma. Studies have shown that toluene does not cause adverse effects on the liver, kidney, lungs, and heart.

In contrast to benzene, exposure to toluene over long periods causes general malaise and CNS effect but does not result in hematopoietic injury, cancer, or bone marrow damage. In studies on experimental animals no chronic toxicity or oncogenicity was observed after exposure to 300 ppm toluene for 24 months.

Workers who had been exposed to toluene vapors over a long term did not suffer from clinically significant adverse effects on the CNS.

Xylene (*Dimethylbenzene*). Xylene exists is three isomeric forms: ortho (1,2-dimethylbenzene), meta (1,3-dimethylbenzene), and para (1,4-dimethylbenzene). All three isomers are found in xylenes. Xylene is used in the production of other chemicals including solvents for inks, resins, adhesives, and degreasers and as a fuel component.

Studies in animals have shown that xylene is rapidly distributed to the tissues after absorption with the greatest concentration in the kidney, subcutaneous fat, nerves, liver, lungs brain, muscle, and spleen. Xylene is absorbed rapidly by inhalation, with 60 to 65% of xylene being absorbed. Xylene is absorbed after topical exposure through the skin at a rate of 4.5 to 9.6 $mg/cm^2/h$. Of the inhaled xylene, 90% is metabolized to methyl benzoic acid (toluic acid) and excreted as methyl hippuric acid. Approximately 5% of xylene absorbed through the lungs is distributed to the body fat in man. Xylene accumulates in the adipose tissue and is slowly excreted. Less than 5% of the xylene is excreted unchanged in the lungs. Xylene has a half-life of 20 to 30 h and binds tightly to proteins.

Xylene is very similar to toluene in its toxicological effects, in that it is a primary skin irritant, and like other organic solvents, it can produce defatting dermatitis upon prolonged exposure. Xylene is an anesthetic at concentration above 5000 ppm. Xylene also depresses the CNS with symptoms of fatigue, nausea, headache, and ataxia. Exposure to high concentrations of xylene results in confusion, respiratory depression, and coma.

Alcohols

Alcohols are rapidly absorbed through the lungs, gastrointestinal tract, and skin and distributed throughout the body tissues and fluids.

The enzymes, alcohol and aldehyde dehydrogenases, are involved in the metabolism of ethyl and methyl alcohols. Metabolism occurs primarily in the liver. The rate of metabolism of ethyl alcohol is constant with time and follows zero order kinetics.

Alcohol at sufficiently high doses are recognized depressants of the CNS. Ethyl alcohol is not a significant CNS depressant until a blood level of 30 to 50 mg% is attained. Inhalation of sufficiently concentrated vapours of ethyl or methyl alcohol can result in significant effects on the CNS. These effects are seen at concentration as high as 5000 to 10,000 ppm of ethyl alcohol.

Methyl Alcohol. Methyl alcohol is used widely as solvent in lacquers, paints, varnishes, and in the manufacture of other chemicals. It is absorbed by all routes and is distributed throughout the body proportional to the distribution of body water.

Effects of methyl alcohol upon the immune system and the possibility of carcinogenicity are of particular interest. Acute exposure to methyl alcohol results in depression of the CNS. The symptoms include headache, vertigo, severe upper abdominal pain, blurred vision, bradycardia, slow shallow breathing, coma, and death. Visual disturbances, for example, blurring and loss of acuity, are characteristics of methanol poisoning. The effect upon the eyes has been attributed to optic neuritis which are regresses. However, if the optic nerve atrophies, permanent blindness will result.

Methyl alcohol is metabolized in the liver by alcohol dehydrogenase to formaldehyde and then to formic acid. Most probably formaldehyde is responsible for damage to the optic nerve which may result in blindness following ingestion of large amounts of methanol.

Ethyl Alcohol. Ethyl alcohol is absorbed through all routes of exposure. It is distributed throughout the body according to body water content transferred across the blood brain barrier.

Most ethyl alcohol used as a solvent is denatured by the addition of other chemicals, i.e., methyl alcohol to make it unsuitable for human consumption. Ethyl alcohol is used as a solvent in shellacs and varnishes and as a solvent for perfumes, mouthwashes, and liniments.

Ninety percent of the absorbed ethyl alcohol is metabolized to acetaldehyde, then to acetic acid, and then via the glycolytic pathway to carbon dioxide and water. The remainder is eliminated by exhalation or in the urine, saliva, and perspiration.

The major toxicological effects of ethyl alcohol is on the developing fetus. The placenta is permeable to ethyl alcohol, and it diffuses into the fetal circulation resulting in fetal alcohol syndrome. Repeated ingestion of large quantities of ethyl alcohol can result in numerous toxic effects including hepatic toxicity (cirrhosis) and cardiac toxicity.

Headache and irritation of the eyes, nose and throat occur at exposures above 1000 ppm. Prolonged exposure to low concentration results in drowsiness, lassitude, and inability to concentrate. Sufficient exposure to ethyl alcohol fumes results in inebriation: talkativeness, alterations in behavior, delayed reflexes and reaction time, loss of muscular coordination, and diplopia.

Ketones

Ketones are a group of compound having the formula R_2CO

$$\begin{matrix} R \searrow \\ \quad C{=}O \\ R \nearrow \end{matrix}$$

Ketones are used as solvents in plastics, paints, and dyes and in the chemical industry as chemical reactants in the synthesis of other chemicals.

Three of the most commonly used ketone are dimethyl ketone (acetone), methyl ethyl ketone (MEK), and methyl-n-butyl ketone (MBK). Ketones are rapidly absorbed from the gastrointestinal tract, lungs, and skin. Acetone is excreted in the urine unchanged. MBK is metabolized to carbon dioxide and 2,5-hexanedione.

The ketones cause depression of the CNS when given in high concentrations and can be irritating to the eyes and mucous mem brane. MBK is exceptional toxicologically because it is the only ketone used singly that has been definitely associated with peripheral neuropathies. This "hexacarbon" neuropathy is attributed to the primary metabolite to MBK which is 2,5-hexanedione, a recognized neurotoxic agent and also a metabolite of *n*-hexane.

Chlorinated Hydrocarbons

Only aliphatic chlorinated hydrocarbon solvents are included. Aromatic chlorinated hydrocarbon have been dealt with separately. Although there are many aliphatic chlorinated hydrocarbon chemicals used as solvents, the most frequently used are described in the following text.

Chloroform ($CHCl_3$, Trichloromethane). Chloroform became known not for its properties as a solvent but as an anesthetic agent. Although used widely in medicine, it toxic effects on the liver the heart resulted in disuse. Chloroform is classified by IARC as a suspected carcinogen in animals and by the EPA as a hazardous substance, a hazardous waste, and a priority pollutant.

Chloroform is widely used as a solvent, particularly in lacquers, and in the manufacture of plastic and fluorocarbons (refrigerants).

Exposure to low levels of chloroform in drinking water is common, due to its formation during the chlorination of organic chemicals in water. This is particularly important to the public as water is consumed and used continuously on a daily basis.

Chloroform is absorbed by inhalation, ingestion, and the dermal route. The first products of chloroform metabolism are carbon dioxide and chlorine. It is of interest that carbon tetrachloride is metabolized to a free radical (CC_3–) but chloroform does not form this free radical. Thus, as expected it is not as hepatotoxic as carbon tetrachloride. However, exposure to chloroform does induce centrizonal necrosis and steatosis. Renal and myocardial damage has been noted.

Although chloroform is irritating to the skin and eyes, inhalation in very high concentrations (389 ppm) can be tolerated without complaint.

Carbon Tetrachloride (CCI_4, Tetrachloromethane): Carbon tetrachloride has been used also as an anesthetic and as an anthelmintic agents. It also was used widely as a dry cleaning and degreasing agent and as a solvent for oil, grease, fats, and waxes.

In recent years, increasing analyses of water supplies have revealed significant concentrations of carbon tetrachloride. It is estimated that 20 million people ingest carbon tetrachloride in contaminated water. Emissions of carbon tetrachloride into the atmosphere have resulted in the exposure of another segment of the population.

Acute exposure to high concentration of carbon tetrachloride results in CNS depression which is exhibited as dizziness, vertigo, headache, mental confusion, and loss of consciousness. Also, some exposed person show gastrointestinal effects such as nausea, vomiting, abdominal pain, and diarrhea. Liver and kidney damage may also occur with acute exposures, but these effects are usually seen in chronic exposure.

Methylene Chloride (CH_2 CL_2 Dichloromethane). Methylene chloride is used as a solvent for oils, fats, and waxes and is also widely used as an aerosol propellant, paint remover, and degreaser.

Methylene chloride is considered to the last toxic of the methene-derived chlorinated compounds. After inhalation, it is absorbed into the circulating blood (55%, human 35%), but it is not absorbed through the skin in sufficient quantities to result in systematic effects.

About 5% of methylene chloride absorbed by inhalation is exhaled unchanged while 25 to 34% is metabolized to carbon monoxide. Carboxyhemoglobin blood levels can reach significant levels of saturation after a severe exposure to methylene chloride.

Methylene chloride is irritating to the skin, eyes, and upper respiratory tract, particularly upon repeated exposure. It produces depression of the CNS with the symptoms produced being directly proportional to its concentration and exposure time. Early symptoms include light-headedness.

Toxic encephalography, pulmonary edema, coma, and death resulted following severe exposure to methylene chloride. Although pulmonary edema is a direct effect of methylene chloride, phosgene (CH_3OCI), an extremely toxic gas, is produce upon combustion of methylene chloride.

1,1,1,-Trichloroethane (Methylchloroform). 1,1,1-Trichloro ethane from 1,1,2-trichloroethane and 1,1,2,2-tetrachloroethane which are more toxic chemicals. 1,1,1-Trichloroethane is classified as a hazardous waste and priority pollutant by the EPA. 1,1,1-Trichloroethane is not a proven carcinogen. Solvents such as 1,2-dichloroethane (ethylene dichloride), 1,1,2-trichloroethane, 1,1,2,2-tetrachloroethane, and hexachloroethane,(p-dichloroethane) are classified as carcinogens by the EPA since they have produced tumors in mice.

1,1,1-Trichloroethane is used primarily as a solvent, degreaser, dry cleaning agent, and propellant.

1,1,1-Trichloroethane is rapidly absorbed from the lungs and gastrointestinal tract. Most of the absorbed 1,1,1-trichloroethane is exhaled by the lungs unchanged. Small amounts are metabolized to trichloroacetic acid and trichloroethanol.

The principal organ affected by 1,1,1-trichloroethane is the CNS with symptoms of depression ranging from headache, disorientation, and drowsiness to convulsions, stupor, coma, and death.

Tetrachloroethylene (Perchloroethylene, PCE, $Cl_2C=CCl_2$). Tetrachloroethylene is widely used in commercial dry cleaning solvents, degreasers, fumigants for grain, and as a veterinary anthelmintic. It is classified as a hazardous waste and priority pollutant by the EPA.

Following inhalation and absorption, tetrachloroethylene is distributed throughout the body and stored in fatty tissues. Absorption through the skin may occur to some extent. Most of the tetrachloroethylene is exhaled through the lungs, but an appreciable amount is metabolized to trichloroacetic acid.

Exposure to tetrachloroethylene may result in depression of the CNS which is exhibited as light-headedness and confusion, disorientation, coma and death.

5

Soil Toxicology

Introduction

Soil is the ultimate receptor for most pollutants released into the environment. Chemicals adsorbed to particulate matter that is released into the atmosphere are eventually deposited onto soils or surface water. In surface water, particulate matter will ultimately settle onto sediments (i.e. water-saturated soil). Accidental spills, pipeline leaks or leaking underground storage tanks may release pure chemicals to the surface or subsurface soil. In addition, land disposal has been the most common method of for centuries, mainly due to economic reasons, for most of our municipal waste and much of our industrial waste. Land disposal units include pits, ponds, lagoons, landfills, septic lines, land treatment facilities, and open dumps. Although these disposal practices differ greatly, yet each procedure may result in the placement of hazardous constituents in direct contact with the soil. Under proper management, these procedures take advantage of the tremendous capacity of the soil to retain and degrade toxic chemicals.

It is extremely rare for soil contamination to consist of a single pure chemical. When an accidental spill does release a pure product to the soil, adsorption and degradation may rapidly change the composition of the spilled product. In most cases, soil contamination consists of a complex mixture of organic and inorganic chemicals. The constituent composition and concentration of this mixture will exert a significant influence on the ultimate environmental fate of the toxic constituents. Conjunctively, the environmental fate of hazardous constituents will exert a significant influence on their toxicological properties. The following text we will consider the influence of physical and chemical properties on the environment fate of several classes of organic and inorganic chemicals, as well as some of the interactions of complex mixtures of chemicals in the soil.

ORGANIC CHEMICALS IN THE SOIL ENVIRONMENT

The Soil Medium

Soil is unique in the sense that it is the dynamic natural body that forms, exists, and acts at the interface of the lithosphere, atmosphere, and biosphere. It is a medium in which many chemical, biochemical, biological, and microbiological processes occur. These processes may immobilize, transform, or degrade organic chemicals in the soil.

The micro and macro properties of soil develop over time due to the combined effects of climate, biotic activity, and topography acting on the parent material. By volume, soil is about 45% minerals, 5% organic matter, 20 to 30% air, and 20 to 30% water.

The primary particles in soil, those that resist further breakdown, are sand, silt, and clay (Table 5.1). The relative proportions or size distribution of these primary particles is known as the soil texture. When soil primary particles flocculate or are cemented together into three dimensional secondary particles or aggregates this is referred to as soil structure. In fine textured soils, such as silts and clay, structural development can cause natural macropores to develop between the planes of the aggregates which may allow dissolved contaminants or spilled chemicals to bypass most of the soil particles and move rapidly and extensively.

Human exposure to organic chemicals released from the soil environment can occur by several pathways:

1) direct contact with the contaminated soil,

2) inhalation of dust or volatilizing chemicals

3) ingestion of soil

Table 5.1 : Characteristics of Primary Soil Particles

Particle	*Size*	*Main Elemental Composition*	*Surface Area (m^2/g)*	*CEC (meq/ 100 g)*	*Chemical Reactivity*
Sand	0.05–2.0	SiO_2	0.1	2	Very active
Silt	0.50–0.002	SiO_2	1.0	9	Inactive
Clay	<0.002				
Kaolinite		Al,Si, Fe, Mg	5–20	3–15	Low activity
Mica		Al,Si,Fe. Mg, K	~100	15–40	Active
Montmorillonite		Al,Si, Fe,Mg	~750	80–100	Very active

4) contact with soil runoff water, and
5) ingestion of food crops or drinking water contaminated from soil pollutants.

The soil parameters and chemical properties that control the dissipation of a contaminant between the different environment compartments, including air, water, soil, and biota are described later.

Water Solubility

The water solubility (S_W) of an organic chemicals is a critical property which affect its environment fate and transport in soils. Some organic chemicals exist in the soil pore water as ionic species (cationic or anionic). For ionic species, their solubility in the soil pore water depends on the pH of the soil solution. Other organic chemicals (e.g., many organic solvents and hydrocarbons) do not exist in the soil pore water as ionic species but as neutral solutes with different degrees of polarity. Those chemicals that are polar are electrostatically attracted to the highly polar water molecules and will remain in the soil solution (highly water soluble). Nonpolar organic chemicals have no electrostatic attraction to the dipolar water molecules, then they will lessen their contact with water (hydrophobic) and thus partition out of the soil pore water onto the soil organic matter or clay. Nonpolar organic chemicals may also partition into the soil gas phase and volatilize from the soil.

Soil Adsorbents

Adsorption of organic chemicals by soil particles is an important mechanism for their removal from the soil solution and for the inhibition of their leaching to the groundwater or volatilization from the soil surface. In soil, clay and organic matter are the predominant adsorbents while silt and sand play a negligible role.

Clays are composed of aluminum octahedral and silica tetrahedral sheets bound together by shared oxygen atoms to form 1:1 or 2:1 layers, and are thus commonly known as alumino-silicates. As clays form, silicon atoms (Si^{4+}) are replaced to varying degree by aluminum (Al^{3+}) or Al^{3+} atoms may be replaced by Mg^{2+}, Zn^{2+}, or Fe^{2+} atoms. This substitution leaves unsatisfied negative charges in the layers from the oxygen atoms (O^{4-}). As a result, clay has a net negative charge, and thus the capacity (i.e., cation exchange capacity [CEC] to adsorb and immobilize cationic species or repulse and enhance the mobility of anionic species. In addition, clay can have a very high specific surface area which also affects the physical sorption of organic chemical. Because field soils are a mixture of minerals, their surface area and net negative charge will be proportional to the amount and type of clay present.

The general types of organic materials that have been identified in fractionated soil organic matter are humic and nonhumic substances. Humic substance (well decomposed materials) make up about 85 to 90% of soil organic matter. Humic substances consist of aromatic polymers with aliphatic side chains and carboxyl, phenolic, and alcoholic hydroxyl, carbonyl, methoxy, and nitrogenated functional groups high concentrations of transition metal ions (Fe + Ni > Cu > Zn > Co > Mn > Crl) have also been observed in the humic fraction, suggesting that these metals are an integral part of the humic complex. Nonhumic substances (recognizable plant debris) are composed of complex, heterogeneous aliphatic structures (e.g., fats, waxes, lignins, terpenes, sterols, etc.) and make up about 10 to 15% of the soil organic matter.

Studies of the structure of humic substance indicates they are not composed of condensed, rigid molecules, but rather of flexible, open, random coils capable of molecular expansion which have many charged sites along the length of the polymer. Functional groups containing oxygen are mainly responsible for imparting the highly negative charged (CEC = 150 to 300 meq/100g) acidic nature to the humic fraction. Due to the presence of aliphatic and aromatic groups, soil organic matter can be have both a nonpolar and a polar nature.

Partitioning

The fate of an organic chemical in soil depends upon its distribution or partitioning between the soil and the other environmental compartments—air, water, and biota. Several equilibrium expressions or partition coefficients have been developed to describe the tendency of an organic chemical to migrate from one environment compartment to another. Some of these are given in Table 5.2.

Soil— Soil/Water Distribution

The *soil sorption constant* (K_d) is the basic expression describing the partitioning of an organic chemical between the soil and soil pore water (Table 5.2). K_d is the same constant as defined by the Freundlich adsorption equation. K_d relates the amount of chemical sorbed to soil or sediment to the amount in the soil pore water at equilibrium. K_d is soil specific and must be determined for each soil material. Typically, either batch equilibrium (mechanical mixing of the soil, water, and chemical) or column displacement (chemical transport through saturated soil columns (methods are used to determine K_d.

Soil Organic Matter—Soil/Water Distribution

The primary adsorbent of organic chemicals in soil is the organic fraction. Therefore, to describe the partitioning character-

Table 5.2 : Partition Coefficients Used to Estimate the Distribution of Organic Chemicals between Soil and Other Environmental Compartments

Partition Coefficient	*Equilibrium Definition*
K_d	(μg chem in soil/g soil)/(μg chem in water/g water)
K_{OC}	(μg chem/g soil oc)/(μg chem/g water)
K_{OW}	(μg chem/ml *n*-octanol)/(μg chem/ml water)
K_W	(μg chem/cm³ water)/(μg chem/cm³ air)

istics of a chemical between the soil pore water and soil organic fraction, the *organic carbon* (oc) *normalized soil sorption coefficient* (K_{OC}) was developed (Table 5.2). K_{OC} is essentially independent of the soil mineral properties and express organic chemical sorption on the basis of the organic carbon content of the soil.

Empirical relationships were developed which show that K_{OC} has a highly correlated linear relationship with the water solubility (S_W) of the chemical.

$$\log K_{OC} = 3.95 - 0.62 \log S_W$$

where S_W is in units of mg/l.

The partitioning of an organic chemical between soil matter and soil pore can be estimated by its partitioning between water and an immiscible organic solvent. Octanol, having a relatively low water solubility (~300 to 540), best imitates the soil organic matter and thus the octanol/water (ow) partition coefficient (K_{OW}) has been developed (Table 5.2)

Volatilization from the Soil Surface

Organic vapors are capable of moving considerable distances both vertically and horizontally is soil. Exposure to chemicals volatilized from the soil surface can occur from

1) vapor loss from land disposal facilities such as surface impoundments, landfills or land treatment units,
2) ingestion of plants contaminated by volatilized organics
3) vapor loss from chemical spills on the soil surface
4) vapor loss from underground storage tanks and pipes and
5) vapor contamination of groundwater from volatile organic chemicals migrating out from landfills or subsurface reservoirs of chemicals.

Volatilization of chemicals from soil involves the desorption of the chemicals from soil, movement to the soil surface, and vaporization into the atmosphere. The factors that influence the rate of volatilization include the vapor pressure of the chemical in soil, the chemicals rate of movement to the soil surface, the interaction of the chemical with the soil due to adsorption, and the concentration of the chemicals in the soil with depth. Soil properties such as porosity, bulk density, water content, organic matter, and clay content, as well as atmospheric conditions such as air temperature, air flow, and turbulence will also influence volatilization.

Soil Runoff and Erosion

Organic chemicals may be released from soils into surface waters by dissolution into unoff water or by adsorption onto soil particulate matter that is lost by erosion. Chemical losses by these pathways can occur from both agricultural and industrial activities. Industrial activities may also contribute to chemical contamination by erosion or runoff. Runoff water from soils amended with hazardous industrial wastes have been shown to contain significant amounts of mutagenic constituents.

Plant Uptake

Major sources of soil pollutants which may contaminate food crops and forages include pesticides and industrial or municipal waste disposed on agricultural land. Plant uptake of organic chemicals is a complex phenomenon which is influenced uptake are time related and include the chemical half-life and the growth period of the plant. Chemicals with a half-life $T_{1/2}$ of <10 d have low potential for plant uptake. As the chemical $T_{1/2}$ becomes longer, and the growth period of the plant increases, the potential for plant uptake of the chemical also increases.

Environment factors, such as ambient temperature soil water content, pH, and organic carbon content, will also effect plant uptake. For example, the phytotoxicity of some pesticides increases with increasing ambient temperature and soil water content but decreases with increased soil organic carbon content. Plant uptake of the pesticide chlorsulfuron is twice as much when the soil pH is 5.9 than when the pH is 7.5 For this acidic chemical, plant uptake is controlled by its polarity and dissociation constant.

The basic chemical properties which influence the behavior of organic chemicals and their "availability" to the plant include vapour pressure K_{OW}, S_W. The effect of these coefficients on plant uptake depends on the mechanism of uptake. There are four possible mechanisms of plant uptake.

1) root uptake and transport and to the higher plant parts,
2) vegetative uptake of chemical vapors from the air.
3) external contamination by solids (i.e., soil or dust) followed by penetration into the plant and
4) transport through oil cells in oil-containing plants.

Chemical penetration of the plant will usually proceed through at least two of these pathways, but uptake and vegetative uptake or organic vapors are often the main routes of plant contamination.

INORGANIC CHEMICALS IN THE SOIL ENVIRONMENT

The major categories of inorganic pollutants include heavy metals, nitrogen, phosphorous, some acids and bases, salts, and halides. Source of common inorganic soil contaminants are given in Table 5.3.

Metals

The categories of heavy metals that the environmentally important soil pollutants include:

1) carcinogenic metals (e.g. arsenic, chromium (Chromate), beryllium, and nickel are known carcinogens,
2) metals which are mobile in soil at neutral pH, and
3) metals which move through the food chain. The levels of heavy metals naturally occurring in soils and those levels which are recommended in soils and drinking water by regulatory agencies are provided in Table 5.4.

Coarse textured soils or fine extruded soils with macropores may enhance the mobility, thus leaching metals into permeable subsoils or aquifers. Other factors that affect the migration of metals in soils include organic matter content, Cation Exchange Capacity (CEC), soil, moisture, and pH. Of these factors, pH and CEC will usually predominate. Under soil conditions of pH$\geq$7, molybdenum and selenium can leach from soil.

Plants have a wide range of tolerance to heavy metals, and the mechanism for this tolerance varies between species. Some plants exclude metals uptake at the root soil interface. Others can chelate the metal once it is within the root. Still other plants (accumulators plants) may take up high levels of metals with no apparent deleterious effect because they have mechanisms to prevent metals from reaching sensitive plant parts. In some cases accumulator plants could be used to lower metal concentrations in soil.

For arsenic, copper, nickel, and zinc the food chain is protected because the toxic concentration of these metals in plants is less than that for animals. On the other hand, cadmium, selenium, and molybdenum are not toxic to plants at high concentrations and can be accumulated in plants at levels that may be toxic to animals. Other metals such as lead, cobalt, and mercury, can also enter the food chain via plant uptake, but to a lesser extent.

Table 5.3 : Common Inorganic Soil Contaminants

Contaminant	Source
Aluminum	Paper coating pretreatment sludge and drinking sludge
Antimony	Paint formulation, textile mills, and organic chemicals producers.
Arsenic	Production of pesticides and veterinary pharmaceuticals, and wood preservatives.
Barium	Manufacturing plants
Beryllium	Smelting industries and atomic energy projects
Boron	Decomposition of organics
Bromide	Agricultural fumigants, industrial wastes, photographic supplies, and pharmaceuticals
Cadmium	Cd-nickel battery production, pigments for plastics and enamels, fumicides, and electroplating and metal coatings.
Cesium	thermionic power conversion and ion propulsion research, nuclear fallout
Chloride	Chlorinated hydrocarbon and chlorine gas production
Chromium	Corrosion inhibitor, dyeing and tanning industries, plating operations, alloys, antiseptics, defoliants, and photographic emulsions.
Cobalt	Steel and alloy production, paint and varnish drying agent and pigment and glass manufacturing.
Copper	Textile mills, cosmetics manufacturing, and hardboard production sludge.

Contd.

Table 5.3 : Contd.

Contaminant	Source
Fluoride	Phosphatic fertilizer, hydrogen fluoride and fluorinated hydrocarbon productions, and refinery waste.
Gallium	Smelter or coal processing plants.
Gold	Medical isotopic waste
Iodide	Pharmaceutical and analytical chemistry wastes
Lead	Pb battery manufacture, fuel additives, manufacturing of ammunition, caulking compounds, solders, pigments, paints, herbicides, and insecticides
Lithium	Carbonates of calcareous parent material
Manganese	Iron and steel industries, disinfectants, paint, and fertilizers
Mercury	Electrical apparatus manufacture, electrolytic production of CI and caustic soda, pharmaceuticals, paint, plastic, paper products, Hg batteries, pesticides, and burning of coal and oil
Molybdenum	Steel and alloy production, pigment, filament, lamp and electronic tube production
Nickel,	Production of stainless steel, alloys, storage batteries, spark plugs, magnets, and machinery
Nitrogen	Sewage sludge, wastewaters, and animal wastes
Palladium	Platinum extraction, dental alloys, electrical contracts, and jewelry
Phosphorous	P quarries, fertilizer, and pesticides
Radium	Uranium processing
Rubidium	Superphosphate fertilizer and coal
Selenium	Coal power plant fly ash
Silver	Photographic, electroplating, and mirror industries
Strontium	Atomic fallout
	Kraft mills, sugar, refining, petroleum refining, and copper and iron extraction
Thallium	Fertilizer and pesticide manufacture, sulfur and iron refining, and cadmium and zinc processing.
Tin	Tin can production
Tungsten	Nuclear waste
Uranium	Radioactive waste
Vanadium	Steel and non-ferrous alloys
Zinc	Brass and bronze alloys production, galvanized metal poduction, pesticides, and ink.

Table 5.4 : Comparison of Levels of Heavy Metals Naturally Occurring in Soils and Levels Recommended by Regulatory Agencies for Soil and Drinking Water

Contaminant	*Common Range (ppm)*	*Average Level (ppm)*	*Regulatory Level (ppm)*	*Drinking Water (ppm)*
Ag	0.01 – 5.0	0.05	5.0	0.05
As	1 – 50	5.00	5.0	0.05
Ba	100 – 3000	430.00	100.0	1.00
Cd	0.01 – 0.7	0.06	1.0	0.01
Cr	1 – 1000	100.00	5.0	0.05
Hg	0.01 – 0.3	0.03	0.2	NA
Pb	2 – 200	10.00	5.0	0.05
Se	0.1 – 2	0.30	1.0	0.01

Plant Nutrients

Nutrients essential to plant growth may be a hazard the environment if they occur in excess quantities or in unsuitable forms. These elements may be harmful to both plants and mammals. The main plant nutrients than can be soil pollutants are nitrogen and phosphorous.

Inorganic nitrogen (N) occurs in soils in various states and such as ammonium ion (NH_4^+) ammonia (NH_3) molecular nitrogen (N_2) nitrate (NO_3^-) and nitrite (NO_2^-). Ammonia, a gas, can be produced if ammonium is exposed to a high pH. Ammonium may also be adsorbed to soil particles at cation exchange sites due to its positive charge. Nitrate and nitrite are anions which are repelled by soil and thus are readily leachable. Although nitrite is toxic to plants in low concentrations, nitrates can be used by plants, and microorganisms.

Wastes applied to soils that are high in nitrogen should be analyzed to determine their form(s) of nitrogen. Nitrogen transformation or plant uptake, leaching, and volatilization can affect the various forms of nitrogen in soil. The form of nitrogen ion of major concern is the nitrate ion. This is because of its high mobility in the soil and water. Also, microbes can convert nitrate to nitrite which can cause methemoglobinemia. The maximum allowable level of nitrate in drinking water is 10 ppm.

Phosphorous does not represent a direct human health hazard, but it may cause a deterioration of surface water if released from soils. Phosphate (PO_4) is a major cause of eutrophication in lakes and ponds. Phosphorous (P) concentrations typically found in soil range from 0.03 to 3.0 mg/L, and the lower value is usually associated with soils near groundwater. Phosphorous applied to the soil as fertilizer is released primarily via erosion because surface soils tend to strongly adsorb phosphorous. With time, mineralization of soil organic matter results in the release of P which is available for plants, microbes, or leaching. Mineralization rates, soil pH, CEC, clay content, and mineralogical composition will affect the efficiency of the soil for retaining P.

Acids and Bases

Acids and bases disposed of in soil usually should be neutralized (pH=7) before it is mixed into the soil. The capacity of a soil to buffer an acidic or basic waste can influence the loading rate of the waste. Should the soil buffering capacity be exceeded, the soil pH can be adjusted by adding lime or a weak acid. If both acidic and basic waste are to be co-applied, the basic waste should be incorporated into the soil first, then the acidic waste can be applied. If this sequence is not followed, the soil may partially dissolve and metals in the soil minerals may leach from the soil. Before any waste with an extreme pH is applied to soil, bench scale tests should be conducted to determine the effect of the waste on the physical and chemical properties of the soil.

Salts

Salts include many substances the produce ions (other than hydrogen or hydroxyl ions) when dissolved. Salts commonly found in soils include calcium, magnesium, sodium, potassium, chloride, sulfate, bicarbonate, and sometimes nitrate. Salt concentrations in soil may increase due to fertilizer or waste applications, soil moisture evaporation, or irrigation.

The salinity of a soil is determined by measuring the electrical conductivity of the soil water (EC), total dissolved solids (TDS), osmotic pressure, percent salt by weight, or the normality. Elevated levels of salt affect plant growth and the physical structure of soil. Elevated levels of sodium in soil can cause soil clay to disperse, thus causing the soil to become impermeable to water movement both into and through the soil. In fine textured soil, inadequate drainage may prevent the leaching of the salts and thus increase the salinity of the soil. Salts can be sorbed by soil colloids or precipitate as insoluble compound, or they may be soluble and leach from the soil or leave the soil in surface runoff. Excess salts may be removed from soil by leaching it with solutions of $CaSO_4$.

Halides

Highly reactive halogens have stable anions that are referred to as halides. These include fluoride (F^-), chloride (Cl^-) bromide (Br^-), and iodide (I^-). While halides naturally occur in soils, excessive levels may threaten the well being of animals, cover crops, and microbes. Cl^- is essential to both plants and animals while I^-, and perhaps, F^-, is essential to only animals.

F^- occurs naturally in soils at concentrations ranging from 30 to 990 mg/kg. F^- is mobile in soil, but application of lime will decrease its plant uptake and leachability, F^- levels are monitored due to its health hazard in drinking water. The allowable level of F^- in drinking water is dependent on the annual average maximum daily air temperature. With air temperature $\leq$12°C, the maximum F^- level can be 2.4 mg/L, and with air temperature from 26.3 to 32.5 C the maximum F level is 1.4 mg/L.

Cl^- is the common in almost every waste stream as either a by-product or as a contaminant from the water source. Soil concentrations of Cl^- vary considerably but average about 100 mg/kg. Most forms of Cl^- are soluble, hence, Cl^- is readily leachable. Soil Cl^- should be monitored by measuring the plant uptake and the concentration in soil leachate. Cl^- should not exceed the 250 mg/l drinking water standard in the soil leachate.

Br^- concentrations in the soil typically range from 2 to 100 mg/kg. Br^- naturally occurs as the bromide ion but can be found in smaller concentrations as bromate (BrO_3^-) and bromic acid. Br^- salts of calcium, magnesium, sodium, and potassium are readily leachable from soils. Therefore, the Br^- that is found in soils is usually that chelated by organic matter. Before a waste containing Br^- is applied to a soil, the natural soil Br^- concentration, plant uptake, and Br^- leachability should be considered to maintain its concentration below phytotoxic levels.

The concentration of I^- in soils ranges from 0.1 to 10 mg/kg. Since it is only slightly water soluble, I^- is mainly retained in soil because it is chelated by soil organic matter or forms insoluble precipitates with phosphate and sulfate minerals. Animal uptake of I^- can be controlled via plant uptake even though I^- is not essential to plants. Phytotoxic levels of I^- in plants (5kg/kg) may result from excess salts in the soil. Waste application should not cause the concentration of I^- in the soil to reach phytotoxic levels.

REFERENCES

Brady, N.S. *The Nature and Properties of Soils*. Macmillan Publishing Co. Inc. New York, 1974.

Greenland, D.J. and M.H.M. Hayes (Eds.) *The Chemistry of Soil Constituents*, John Wiley & Sons Inc. New York, 1978.

Thibodeaun, L.J. *Chemodynamics, Environmental Movement of Chemicals in Air, Water and Soil*. John Wiley & Sons. Inc. New York, 1979.

6

Toxic Metals in the Environment

INTRODUCTION

Over 40 elements in the environment are classified as metals. Many, such as the alkaline earth group and some trace elements, are essential for life; other have great potential for toxicity. Macronutrients such as calcium, magnesium, iron, potassium, and sodium are particularly important in sustaining life but may become toxic in excessive concentrations. Trace elements such as chromium, cobalt, copper, manganese, nickel, selenium, and zinc are structurally part of important molecules and may serve as cofactors of enzymes in metabolic process. Excessive concentrations of these elements are also toxic harmful. Some elements such as lead, cadmium, and mercury have harmful effects on biological tissues at any concentration.

In determining hazards associated with heavy metal contamination of the environment, it is most important to determine the most sensitive environmental organism. Animals usually are more sensitive to heavy metal contamination than are plants, and humans are universally considered as the most sensitive, as well as the important target species.

METALS IN THE BIOSPHERE

Air

It is a mixture of nitrogen (78.08%) and oxygen (20.95%) are contains other minor constituents such as argon (0.93%), carbon dioxide (0.03%) and some trace gases (e.g. neon, helium, methane, krypton, nitrous oxide, hydrogen, ozone, xenon etc.) with levels ranging between 0.09 and 18 ppm. Pollutants such as dust, smoke, soot, industrial and automobile exhaust and gaseous and particulate matter, may be present in varying concentrations depending upon population, vehicular density, and location and type of industrial units installed in that area. Metal compounds, though not a normal constituent of air, exist variety of forms particularly in urban areas. Usually, they are in the particulate phase. The sources of metallic contaminations in air are both natural (e.g.) terrestrial, marine, volcanic, biogenic) as well as anthropogenic (e.g. combustion, Industrial automobile). Urban industrial air may contain suspended particulates of a wide variety of materials including heavy metals, such as lead, nickel, zinc, copper, cadmium, etc. depending on the occurrence of specific type of mining, milling or manufacturing units located in that area. Particles of 0.1–10 μm in diameter settle very slowly and remain suspended in the air for a longer duration of time. The small size particles < 1 mm in diameter) of these particulates allows them to be inhaled deeply into the respiratory system. Metal ions are absorbed by the lungs ten times more efficiently than the intestines. They are, therefore, of grave concern as a health hazard specially in densely populated areas and work placed where high levels of metal-containing vapours and particles persist.

Emission of trace metals in the northern hemisphere (80%) is much higher than in the southern hemisphere (30%). A sharp increase in anthropogenic emissions have been observed during the last two decades.

Water

Pure water, which does not contain any soluble impurity does not exist in nature. Water for drinking and domestic purpose is usually procured form rivers, lakes, wells and springs. Such waters may contain organic matter, salts of calcium, iron, magnesium, potassium, and sodium and traces of carbon-dioxide, oxygen, nitrogen, ammonia and other gases from the atmosphere. Presence of potentially hazardous heavy metals in water should be considered abnormal and this usually affects its appearance and palatability. The concentration of these metals in rainwater in urban areas may sometimes exceed safe levels for human consumption. Metals of prime environment concern e.g. cadmium, copper, lead, mercury, zinc, etc.) are present in very low concentration in sea-water, but human influence is increasing these levels rapidly. Fresh-water ecosystems are influenced by air-heavy metals levels and addition from anthropogenic sources. The chemical forms of these metals in water are accessible to the biota through significant accumulation in the food chain. It is important to note that at lower concentration of dissolved metals in natural water, even smaller alteration in aquatic environment may critically influence in the stability of the ecosystem. Besides the absolute concentration, the physico-chemicals state and speciation in an aqueous system should also be looked into for any assessment of heavy metals in water for human use.

Drinking being one of the routes of intake of heavy metals into the human body. It has drawn special attention.

Soil

Man depends on soils to produce his food. For terrestrial ecosystems, it is the point of entry of most materials into living matter. Soil is composed of finely divided rocks mixed with decayed vegetable and animal matters. The element composition of soil is largely dependent on that of the rocks from which it is formed. Wide variation in that soil composition, therefore, occur in different parts of the world. Besides, it is affected by a number of factors, such as weathering, soil-forming processes, climate conditions, geochemical affinities, sewage sludge, waste treatment, discharge from factories fertilizers, pH, and deposition redox potential, amount and kind of organic matter, mining activity, aerial pollutants, etc.

Soil has complex functions which are beneficial to man and other living organisms. It acts as a filter, buffer, storage and transformation system and thus protect the global ecosystem against the adverse effect of environmental pollutants. These function can be performed effectively only if the normal properties of soil are preserved and the natural balance is not unduly disturbed.

Plants

Plants absorb metals from soil, water and air, but the chief source of metal absorption is soil. Uptake from the soil depends on the total content of the respective metal and its accessibility to roots and transfer across the soil-root interphase. The total amount of metal in a soil is affected by the natural resources of the particular area and the agricultural and industrial activities.

Accessibility to roots depends on (a) Soil - type, pH, drainage status, presence of organic matter, sewage sludge and microbial activity, (b) Plant - species, part, season of collection, and (c) Metal/ Metalloid- chemical form, location, etc. Metals present in the ionic form in the soil solution are readily available, while those which are firmly bound to rock minerals are the least available. Accessibility of Co, Mn and Ni increases with decrease in pH, while that of Mo and Se increases with increase in pH. Uptake of metals like Cu is only marginally affected by changes in pH of the soil. Poor drainage increases release of Co. Ni and V as also accumulation of organic matter which influence microbial activity and metal uptake.

Wide variations in metal uptake are observed in different plant species. Some plants are known to have specific affinity for accumulating certain metals. Some examples are

Astragalus sp. (Se)

Crotalaria cobalticola (Co),

Phaseolus vulgaris roots (Zn),

alga *Chlorculla Vulgaris* (Au),

Alyssum sp. and

Sebertia accuminata (Ni) affinity for metal.

Aquatic plants accumulate several hundred fold amounts of Ag as compared to other plants. Some Se-accumulating plants may attain a concentration of upto 1000 to 10,000 ppm of this metalloid. Young leaves and seed-heads reportedly have higher cu and Zn concentrations while roots and fruit peel contain more Pb than other parts of the plant. The soil-root interphase is not mechanical sieve but an active biological process tuned to the metabolic requirements of the plant. Chelation or precipitation near root may, however, affect metal uptake to some extent.

Besides the root, metals may enter the plant through aerial parts including the leaf surface. High Pb content (from motor exhaust fumes) may be observed in plants growing near busy traffic. Minute particles (< 2μm in diameter) of some metals/metalloids, e.g. As, Cd, Cu, Se and Zn, may dissolve in rain water and enter the plant through the leaf.

An interplay of a number of factors is involved in metal-uptake by the plants. This makes it virtually impossible to make general statements for the heavy-metal content of plants. Some metals/ metalloids, however, are considered essential for the normal growth and metabolic functions of plants. These cannot be substituted by other elements in their specific biochemical roles and include Co, Mo (as enzymes cobalamide and nitrate reductase for nitrogen fixation), Cu, Zn (in oxidases and anhydrases for the metabolism of proteins and carbohydrates), Fe (involved in valence changes and photosynthesis), Mn (various enzymes with a role in O_2 photoproduction), Ni (as urease in hydrogenase metabolism) and V (as porphyrins in lipid metabolism).

Plants have a higher capacity to take up metals from soils or atmosphere than their physiological needs. While this may not have any adverse effect on plant itself. It may expose the consumer to higher intake of concerned metals.

Animals

Animals are exposed to metals via food such as plants, other animals, drinking water and through inhalation. The mechanism of metal uptake is usually fast and efficient; The concentration factor (ratio of the metal concentration inside an organism to that outside it) is usually low and depends on a number of factors which include the metal itself, its speciation, the organism and the environmental conditions.

The essential and toxic roles of metals have been studied extensively in recent years. The following metals/ metalloids considered essential for animals and their role is listed in Table 5.1.

Micro-Organisms

Metal uptake by micro-organisms may occur through the

Table 5.1 : Essential Metals and their Role

Metal	*Role*
As	(non-specific growth stimulation)
Co	(constituent of vitamin B_{12})
Cr	(regulator of metabolism of glucose and cholesterol)
Cu	(Constituent of oxidases which are important for the regulation of redox reactions, respiration, cartilage formation)
Fe	(in hemoglobin, Cytochromes, catalase, peroxidases)
Mn	(as arginase, Superoxide dismutase and pyruvate carboxylase in urea synthesis, protecting mitochondria from free radical damage and citric-acid cycle)
Mo	(aldehyde and xanthine oxidases in formation of fatty acids and uric acid respectively)
Ni	(constituent of several enzymes, undefined growth stimulation)
Se	(asglutathione peroxidase, role in tissue oxidation and cardiomyopathy)
Sn	(as a constituent of gastrin with digestive and growth-promoting functions)
V	(in lipid metabolism and bone mineralization and)
Zn	(Constituent of several enzymes : proteases, anhydrases, superoxide dismutase with role in protein biosynthesis, energy metabolism, protection against damage by superoxide radicals, fertility, etc.)

processes involving non-specific absorption and/or metabolically mediated mechanisms. While the latter processes are observed in viable cells only, non-specific binding on the cell surface may occur in both viable and non-viable cells. Metal cations are readily absorbed to anionic ligands (carboxyl, hydroxyl, phosphoryl and sulphydryl groups) at the cell surface through a rapid, reversible process, which is usually independent of temperature/energy metabolism, metabolically mediated mechanisms, chiefly involve divalent and, to a lesser extent, monovalent cations transport systems. Some metals, such as Cd, Co, Mn, Ni, and Zn can replace Mg in several enzymes/

co-enzymes which play a role in phosphate transfer in several bacteria and fungi. Certain organisms (e.g., *Staphylococcus aureus*) Utilize metabolic processes to exclude metal ions from the cell and establish resistance. Alteration of the form of occurrence of metals by micro-organism through chelation, complexion, methylation etc. significantly affect their bioavailability and toxicity. The presence of glutathione, citric acid, oxalic acid, humci acid, methionine and other amino acids may also reduce metal toxicity. An interplay of various factors, like pH, temperature, redox potential, sunlight, composition of water and soil, number and type of organisms, period of exposure, interaction with other metals, etc., is responsible for marked differences between the toxicity of metals to micro-organisms in pure culture and in their natural habitat.

Biomethylation

A number of metals and metalloids may undergo biomethylation leading to their detoxication. On the other hand, many metals or metalloids, instead of being detoxified, become more hazardous to living beings including humans through this process. Micro-organisms play an important role in the environmental cycles of some elements. Organometallic compounds in our environment are often produced by biomethylation where the transfer of a methyl group(s) is mediated biologically from a donor molecule to the inorganic form of the element. Microbes possess efficient mechanisms to carry out such relations. For example, various species of bacteria, fungi and algae are known to biomethylate some metals and metalloids, *viz*, tin, lead, cadmium, mercury, thallium, arsenic, germanium, selenium, tellurium etc. As a result, the methylated forms of arsenic and antimony have been observed in natural waters while methyl mercury compounds are found in the body organs of a number of marine organisms. Bacteria and fungiin sewage sludge may methylate a number of above mentioned elements contaminating the residential surroundings.

SOURCE OF ENVIRONMENTAL METALS

The earth's crust is the one and only source of all metallic elements found in the environment. Metals are neither created not destroyed by human but are redistributed naturally in the environment by both geological and biological cycle. The industrial and technological activities of man simply shortens the ore phase of the metal, forms new metallic compounds, and introduces metals into the atmosphere.

Industry

Combustion of fossil fuel releases about 20 toxicologically important metals, into the environment including arsenic, beryllium, cadmium, lead, and nickel. Industrial products and used industrial material may contain high concentrations of toxic metals. For example, mercury is used by the chlor-alkali industry to produce chlorine and caustic soda in the pulp and paper industry and in the production of battery cells, fluorescent bulbs, electrical switches, paints, agricultural products, dental preparations, and pharmaceuticals.

Although both adults and children may be exposed to lead dust from such industries as battery manufacturing through contact with workers clothing, children are frequently exposed to lead through ingestion of peeling paint chips. The environment release of lead from the combustion of tetraethyl lead containing auto or industrial fuels remains the greatest source of exposure.

Cadmium, a by-product of zinc and lead mining, is an important environmental pollutant. It has many industrial uses in paints, pigments, batteries, and plastics. Another use is an anticorrosive agent for steel, iron, copper, brass, and other alloys.

Agricultural Products

Heavy metals present in fertilizers cd, cr, cu, manganese, molybdenum, nickel and zinc. The major environmental sources of arsenic are pesticides, herbicides, and other agricultural products. Lead arsenate, in addition to being a component of industrial effluents, has been used as an agricultural pesticide Fungicides containing mercury contribute to environmental contamination. Eventually, many of these metals may accumulate in agricultural soils and pose a hazard to plant growth and animal nutrition.

Food and Food Additives

For humans, the ingestion of food and food additives may represent the largest source of exposure of metals. The ingestion of metals normally follows two primary routes : the first involves direct, accidental contact with metal-bearing food or material. Ingestion of foreign substances such as lead-containing paint chips by children is one of the best examples of this method. Since paint chips may contain 50 to 100 mg of lead in a 1 cm paint chip, even a few chips a day may exceed the acceptable daily intake.

The second route of ingestion of metals of through the bioaccumulation of metals in the chain. The well known *Minamata disease* is the result of massive human exposure to methylmercury-contaminated fish. Microorganisms in sediments methylate mercury into methylmercury. Methylmercury can then be assimilated by fish through ingestion of contaminated food sources and accumulation in their tissues. Uptake by fish in Minamata Bay apparently occurred as the industrial release of mercury compounds into the bay preceded the accumulation of mercury by edible fish. Other toxic metals such as cadmium are found in meat, fish, and fruit. Shellfish such as mussels, scalps, and oysters are a major source of dietary cadmium.

Sewage sludge containing nitrogen, phosphorous, and sulphur as well as the toxic metals cadmium, chromium, copper, lead mercury, nickel, and zinc has been applied to agricultural soils to improve crop production. Although leaching of heavy metals through soils is limited, plant uptake may translocate the meals into the food chain, eventually being consumed by humans at the highest tropic level.

TOXIC METALS IN THE ATMOSPHERE

The atmosphere is a dynamic system consisting of hour principal zones, the troposphere, the stratosphere, the mesosphere, and the thermosphere. These four zones are separated by three regions of temperature inversions, called the tropopause, the stratopause, and the mesopause. The zone closest to the earth is the troposphere and is of the greatest concern for the transport of metals.

Temperature and winds are a major influence on the rate and volume of movement of heavy metals, particulate matter, aerosols, or even vapor in the atmosphere. Changes in air temperature are responsible for vertical air movement. Even though the temperature of tropospheric air normally decreases inversely with an increase in altitude, temperature inversions can also occur resulting in horizontal movement and mixing.

An excellent example of the meteorologic effects on long-distance movement and deposition of atmospheric metals was seen following the Chernobyl reactor explosion in 1986. The radioactive clouds move in two separate directions and covered most of Europe of approximately two weeks.

Although radioactive elements from nuclear accidents and explosions are widely distributed, volcanic emissions contain the highest concentration of metals found in any natural contributor of metals into the atmosphere. Antimony, arsenic, cadmium, lead, and selenium are discharged into the atmosphere from volcanic sources while several heavy elements are found in effluents from other geothermal sources. Although the highest concentrations of metals are in volcanic emissions, windblown material is the largest contributor of total heavy metals to the atmosphere with forest fires, vegetation, and sea spray contributing a great amount of lead and cadmium than volcanic emissions.

Anthropogenic sources of metal contamination in the atmosphere are more concentrated in highly industrialized urban area. These source include

(1) industrial sits and engines involved in the combustion of fossil fuels such as coal, oil, and natural gas,

(2) metal manufacturing plants and foundries,

(3) mines and smelts,

(4) refuse incinerators, and

(5) cement production sites.

Atmospheric levels of heavy metals are significantly lower in rural and remote areas and mainly a reflection of emissions from natural sources.

Combustion of either coal or oil results in the discharge of arsenic, bismuth, cadmium, chromium, copper, lead, manganese, mercury, nickel, selenium, and zinc into the atmosphere as global aerosols. Antimony and indium are also found in significant quantities in urban area aerosols.

TOXIC METALS IN WATER AND SEDIMENTS

The hydrosphere accounts for a great area of the earth's surface than the lithosphere and is divided into lakes, rivers, estuaries, and oceans. Metals exist in the hydrosphere as dissolved material and suspended particular and are in deposited sediments. Sediments in rivers, lakes, estuaries, and oceans account for the main sinks of heavy metals in the hydrosphere. In estuaries, heavy metals from the atmosphere and rivers accumulate, thus permitting chemical and physical reactions to occur before being washed out into the ocean.

Atmospheric deposition, soil leaching, runoff, erosion, and breakdown of mineral deposits all contribute to the concentration of metals in natural water supplies. Anthropogenic source include mining, smelting, burning of fossil fuels, leaching from waster dumps, urban, runoff, sewage, effluent, waste dumping, and agricultural runoff.

The levels of trace metals are usually higher in rivers than in ocean because metals from point and nonpoint sources are discharged into the rivers. Changes in concentration of metals in rivers are easily detected because of their rapid rate of transport. Levels of certain trace metals such as cadmium, mercury, and lead are closely correlated with variation in population density along a river and with seasonal variations in flow rate. Concentrations vary inversely with velocity of the river. Likewise, sediment concentrations of a metal in a river varies inversely with distance from the source. Studies of the Rhine River in Western Europe has shown that the cadmium concentrations fall with an increase in velocity of the river in winter and rise with a decrease in velocity in summer and fall.

Levels of heavy metals in estuaries, coastal waters, and sediments vary considerably, depending on inputs. Much of the estuaries is due to atmospheric fallout, and as much as 93% of heavy metal entering an estuary is retained.

The atmosphere is the major contributor of heavy metals to the oceans. This is especially true for lead, with 90% of the lead found in ocean waters derived from the atmosphere. Concentration of heavy metals in ocean sediment vary with geographical locations. Higher levels are often found in coastal waters because of nearby sources of pollution. The profile of heavy metals such as lead in ocean sediment is similar to that found in the water column, with the highest concentrations near the surface..

The effect of metals on aquatic organisms is difficult to determine as many physical and chemical properties such as flow rate contribute to the outcome. Also, the size and nature of the particulates to which the metals are attached affect the toxicity of the metal. In order for a metal to enter an organism it must either be phagocytized or in a solubilized form. Chelation of metals with an organic molecule may alter chemical reactions and membrane permeability, thereby altering adsorption by an organism. Metals may also adsorb to colloidal particles in the water and thus alter the degree of adsorption by the aquatic organisms.

TOXIC METALS IN THE TERRESTRIAL ENVIRONMENT

Many of the heavy metals that occur naturally in the earth's crust are released into the soil through the process of weathering, but the largest natural source of metals to the terrestrial environment if fallout from volcanoes. Anthropogenic sources of metal contamination to the terrestrial environment via the atmosphere included fuel combustion, metal manufacture, foundries, refuse incineration, and cement production. Metals also are introduced into soil as plant nutrients, pesticides, and constituents of waste products. Plant uptake of heavy metals from sewage sludge applications to croplands has been documented.

The predominant metal pollutant found in soil are arsenic, cadmium, lead mercury, and selenium, followed by antimony, bismuth, indium tellurium, and thallium. Levels of metals higher than those shown in Table 5.2 can be found in soils contaminated by mining and agricultural activities and the addition of pesticides

Table 5.2 : Levels of Some Metals in Surface Soils Worldwide

Metal	Symbol*	Range (μg g^{-1})	Mean (μg g^{-1})
Antimony	Sb	0.05 – 260	0.9
Arsenic	As	<0.1 – 97	11.3
Bismuth	Bi	0.13 – 10	0.2
Cadmium	Cd	0.1 – 1.0	0.62
Indium	In	0.7 – 3.0	1.0
Lead	Pb	1 – 888	29.2
Mercury	Hg	0.005 – 1.11	0.098
Selenium	Se	0.005 – 230	0.4
Tellurium	Te	0.5 – 37	---
Thallium	Tl	0.03 – 5.0	0.4

Source : Fergusson, 1990.

and sewage sludge to soil.

Elevated levels of antimony and cadmium may be associated with soil in the vicinity of smelters. High levels of arsenic are associated with soils around metal processing plants and with soils contaminated by arsenical pesticides, sometimes reaching 600 $\mu g\ g^{-1}$ in the latter case. Source of lead contamination include metal working, internal combustion engines using tetraethyl lead in gasoline, paint, and smelting. In some surface soils the concentration of lead can reach as high as 10%. Mercury contamination of soils is usually associated with mining activities, the production of dichlorine and caustic soda, and the agricultural use of mercury compounds such as fungicides.

Heavy metals may enter plants via uptake from the soil through the roots or through foliar uptake. The more common process of incorporation of metals into plants is mainly through the roots by active transport against a concentration gradient. Metals are absorbed by the roots by passive diffusion through cell membranes. Once inside the plant the metal is transported through the xylem to various parts of the plant. The mobility varies from one metal to another, but in general, the levels will be highest in the roots and lowest in the seeds.

The principle route for foliar uptake is through the leaf cuticle. This route of uptake is very significant, especially, for the entry of aerosols containing metals. The route of entry also changes the relative levels of metals found in various structures of the plant.

The concentrations of heavy metals vary widely in the edible parts of the plants, ranging from 0.001 $\mu g\ g^{-1}$ with tin to 20 $\mu g\ g^{-1}$ with lead. Plant species also vary in their uptake of heavy metals.

Although many species of plants are unable to survive on soils containing high concentration of heavy metals, some do survive and flourish. Studies have shown that when the pollutant exerts selection pressures, and if there is suitable range of genetic variation, a considerable number of plant species, especially grasses, can produce genotypes resistant or tolerant to one or more metals. However, apple orchards in the northwestern U.S. become nearly sterile after exposure to heavy metal contamination. If the germination of such an important plant is inhibited at levels below those effecting animal health, the levels of contamination are unacceptable.

Plants growing near metals smelters normally contain high levels of arsenic, which depends on the type of arsenical compound, plant species, soil type, and climate. For instance, vegetables accumulate significant levels of arsenic when grown in soils containing high levels of arsenic trioxide. With increasing levels of lead arsenate, beets, lettuce, and radishes show an increased arsenic uptake while the uptake by broccoli, carrots, eggplants, and tomatoes is unaffected.

Humans may be primary consumers of metal containing plant material, or metals may be transported between trophic levels of a food chain by other means. Terrestrial invertebrates such as earthworms, millipedes, and isopods stimulate decomposition of dead plant material by increasing the surface area for microbial attack. Centipedes and spiders are predators of terrestrial isopods and transfer metals from primary consumers to secondary consumers.

TOXIC EFFECTS OF METALS

The sensitivity of organisms to metal toxicity varies widely with species of plants and animals and genotypes within species (e.g. cultivars of crops) and many factors can modify the response to the toxic dose of metals. Some individuals are genetically adapted to tolerating anomalously high concentrations of certain metals. Homeostatic mechanisms in animals frequently involve special proteins, metallothioneins, which blind with the metals (such as Cd) and render them relatively inactive. Plants have similar compounds and called phytochelatins. It is therefore difficult to generalize about toxicity.

Phytotoxicity. The most toxic metals for both higher plants and several microorganisms are Hg, Cu, Ni, Pb, Co, Cd, and possibly Ag. Be and Sn. Although the occurrence of toxicity will depend on soil factors, such as pH, the plant genotype and the conditions under which it is growing (pot in greenhouses or under field conditions).

The relative toxicity of different elements will be affected by considerable variation of individuals and, in the case of diets, in the composition of the diets. The doses injected into rats and other experimental mammals probably provide a more accurate comparison of toxicity.

ALUMINUM

Aluminium is relatively non-toxic. The fatal doses in animals are, large,–cats, (5, 10g), dogs (35–50g) and cattle (500-g) Harmful effects following exposure include gastro-intestinal disturbances, neutrotoxicity, haematological changes and bone damage.

Aluminium compounds have given rise to encephalopathy in isolated cases in industry and of epidemic proportions in other population groups. Alzheimer's disease, characterized by pre-senile dementia, sclerosis and neurofibril degeneration, is associated with aluminium accumulation in the brain tissue. A cause-effect relationship, however, could not be proved. Current medical opinion does not link its presence as an etiological factor but it is believed that the metal may be a nonspecific marker of degenerating neurons. Dialysis osteodystrophy and dialysis dementia may result from prolonged treatment with water containing more than 50 μg aluminium from natural sources or when aluminium sulphate is added to precipitate particulate matter in domestic water. The clinical manifestations include bone-pain and speech disorder, followed by dementia, convulsions, myoclonus and death. Elimination of aluminum from the dialysis fluid arrests encephalopathy early in the course of a neurological disease. There is no safe level of aluminium in water used to prepare the dialysate and it is absolutely essential to remove it.

Exposure to aluminium dextran and aluminium oxide dusts (particle size : 1-5 micron) has been shown to be carcinogenic in mice. However, epidemiological studies and mutagenic tests in bacteria did not indicate such effects with aluminium compounds.

ANTIMONY

Dermal contact with a high concentration of antimony salts causes irritates of the skin and mucous membranes. Oral intake of excess amounts results ir diarrhoea, vomiting, hepatic damage, metabolic disturbances and circulatory collapse. Inhalation exposure has been reported to cause pneumonitis, fibrosis, bone-marrow-damage and carcinomas. Stibine SbH_3, produces adverse effect on erythrocytes and the nervous system.

A high incident of urinary bladder tumors, seen in patients with schistosomiasis, is suspected to be caused by antimonials. Life-time studies in mice, who were fed 5 ppm antimony potassium tartrate in drinking water, showed no evidence of cancer induction. However, a slightly higher risk of lung cancer was observed among workers exposed to antimony to antimony.

Arsenic

Arsenic is a toxic, non-essential element, and occurs widely in nature. It is used in alloys, pesticides, wood preservatives, and some medical preparations. It was formerly used in plant pigments, but this use discontinued when it was found that, under damp conditions, moulds converted the arsenic the highly toxic gases, a sine, AsH_2, and trimethyl arsine, As $(CH_3)_3$.

It is also present in many sulphide ores of metals and is therefore emitted from metal smelters as an atmospheric pollutants (4-% of the anthropogenic emissions). Coals also contain significant amounts of As and its combustion accounts for 20% of atmospheric emission. Coal ashes are also a significant source of As which can be leached out into waters or the soil. The toxicological importance of As is partly due to its chemical similarity with P which means that As can disrupt metabolic pathways involving P. Both acute and chronic toxicity are recognized and the continual inhalation of airborne forms of As is known to be carcinogenic. Respiratory cancers have occurred in occupationally exposed workers.

The absorption of arsenic depends upon the type of the compound, its solubility and its physical form. Inorganic arsenic is more readily absorbed than organic arsenic and pentavalent from more than the trivalent from. Dermal adsorption, however, is more rapid for trivalent than for pentavalent from. Inhalation usually involves particles containing inorganic arsenic. Most of the ingested (95% of trivalent As and 80% of sea-food AS) or inhaled arsenic is adsorbed through the gastro-intestinal or the respiratory tract.

After absorption, 95-99% of the arsenic binds to globin of haemoglobin in the erythrocytes. Then it is transported to other tissues and body fluids within 24 h. Its deposition occurs in bones, hair, nails and skin where it may persist. Placement transfer of inorganic arsenic has been reported in both humans and experimental animals models. In a study on rats, fetal blood showed levels of arsenic comparable to those of the mother.

Detoxification processes include oxidation of trivalent compounds to pentavalent forms, followed by methylation by methanobacteria forming monomethyl arsenic acid (MMAA) and dimethylarsinic acid (DMAA). In liver, biomethylation occurs mainly by the enzymatic transfer in the methyl groups from S-adenosylmethionine. It has been proposed that the reaction between As^{3+} and Ch_3^- is very unlikely to occur in aqueous solution.

The important examples of interaction of arsenic with other elements include the replacement of phosphorous, negative interaction with iodine and reaction with sulphydryl groups of thiol-containing enzymes. The compounds may also react with thiol groups in the biological systems. Trivalent arsenic compounds may form stable bonds, with thiol groups in the enzymes, disrupting enzyme function. Alkylarsines and dialkylarsines are formed by reductive methylation of inorganic arsenic and methylarsinic acids which easily condense with SH- groups to form alkylbis (organylthio) arsenic. Most (70-80%) of the ingested arsenic is eliminated within a weak, mainly through the urine as MMAA and DMAA. However, other excretory routes include faeces, skin and hair. Urinary analysis serves as a reliable index of exposure to arsenic.

Arsenic is a cumulative poison, causing vomiting and abdominal pains prior to death. It may also cause dermatitis and bronchitis, and may be carcinogenic to tissues to the mouth, oesophagus, larynx and bladder. At the cell level, it an uncouple oxidative phosphorylation and compete with phosphorus in metabolic reactions.

Ingestion of a large dose of soluble arsenic compound, especially on an empty stomach, many cause death within a few hours. The fatal dose of arsenic trioxide in adult humans is 70-180 mg. Usually, swallowing of arsenic is painless but is may cause epigastric pain in some cases. Arsine As H_3, also causes death in 25-30% cases within 24 h of its exposure. The immediate symptoms are severe gastrointestinal irritation, diarrhoea, vomiting, haematuria, leg cramps, shock and coma. Death may result from acute injury to the myocardium with or without brain-stem medullary failure. The surviving patients and cases with chronic exposure to arsenic, show inflammation of the conductive and respiratory mucous membranes, epistaxis, cardiomyopathy, facial oedema, skin diseases such as erythematous or vesicular rashes, scaling pigmentation, hyperkeratosis, transverse striate leukonchia or Aldrich-Ness lines on nails, destruction of red-blood cells and bone-marrow depression (haemolysis, anaemia, leucopenia, jaundice, haemoglobinuria), muscular pain, weakness, renal damage, and symptoms of peripheral neuropathy with both sensory and motor effects such as ataxia, numbness, tingling in limbs, drowsiness, poor memory, episodes of delirium and irrational speech, paralysis, degeneration of inner ear, etc. Arsenicals may also act as skin-contact allergens.

Epidemiological studies have demonstrated a certain association between environmental, occupational, and medicinal exposure of man to inorganic arsenic and cancer of skin and lungs. The use of sodium arsenite as herbicide is forbidden in many countries.

Occupational exposure of workers engaged in the manufacture and spray of arsenical pesticides leads to skin pigmentation, raindrop rash, hyperkeratosis of palms and soles followed by epithelioma of multifocal origin. The black-food disease (dry gangrene) and chronic arsenic poisoning co-occur, indicating a cause-effect relationship between the two. Indicate of squamous cell carcinoma has been reported in some parts of the world consuming water-rich in arsenic.

Acute, subacute or chronic exposure to arsenic (particularly through inhalation) may cause inflammation of the mucous membranes, of nose, laynx and respiratory passages manifested by coryza, tracheobronchitis, cough and perforation of the nasal septum. Increased incidence of respiratory cancers has been reported among occupationally exposed workers of smelting units of arsenic-based industries and pesticides/spray arsenicals.

Arsenic is concentrated by organisms exposed to it and accumulates along food chains. Accumulation in fish seems to be favoured by increasing salinity. Anand (1978) has reported arsenic pollution loads in fish sold in Bombay. Crabs and lobsters have been especially noted for accumulating high concentrations, but no cases of human poisoning seem to have arisen from this.

The role of arsenic in carcinogenesis is controversial. The IARC believes that there is sufficient evidence to show that inorganic arsenic compounds are skin and lung carcinogens in humans. The risk, however, appears to be marginal. A WHO study estimated 1% increase in lung cancer incidence in workers exposed for at least 25 years to 0.25 $\mu g/m^3$ of non-soluble inorganic arsenic, and for the general population, 0.75% pe 1µg/m³.

The carcinogenic effects of arsenic have not been proved. Some reports indicate reduction in the frequency of lung or breast cancer following long-term feeding of arsenicals in experimental animals. People getting arsenical water in USA had a better health record than those getting non-arsenical water. It was further pointed out that if inhalation of arsenic trioxide really caused lung cancer, the disease would have reached epidemic proportions in Prague, where the level of arsenic in air from coal burning is very high. The current opinion is that arsenicals, like many other nutrients, (e.g. vitamin A, selecnium zinc etc.), have synergistic, antagonistic or no effect on carcinogenesis depending upon the form, dosage, and method of administration. The metal may be considered as a promoter, rather than the initiator, of the cancer process.

Barium

The most common naturally occurring form of barium is barium sulphate, sometimes called barite or baryta. Barium derivatives are used as fillers for rubber, linoleum, etc., as pigments in paint, in glass manufacture, in the ceramic industry and in various other industrial applications. The applications constitute the main source of risk, primarily to workers in the appropriate industries. Barium salts are highly toxic. They cause vomiting and diarrhoea, which may be associated with stomach, intestinal and kidney haemorrhage. They also affect the central nervous system, causing convulsions and have been implicated as a cause of pneumoconiosis. Despite the known toxicity of barium salts, barium sulphate is used to cost the alimentary tract for X-ray photographs since barium absorbs X-rays strongly and thus increases the contrast. The insolubility of barium sulphate minimizes its toxicity, and makes its use acceptable.

Beryllium

Beryllium is the most permeable and toxic element of Group II. Beryllium is released during the burning of coal, but the main environmental hazard is to those working in industries where beryllium is produced or used, e.g. the manufacture of nuclear reactors, aircraft and rockets. In humans, beryllium has been shown to damage skin and mucous membranes. It has been shown to cause degenerative lesions of the parenchyma, inhibition of bone metabolism, osteoporosis, impaired hematopoiesis, dermatitis, delayed wound healing, granulomas and lung tumours in experimental animals. Inhalation of soluble compounds of beryllium, dust or aerosols is a serious health hazard. It produces an acute pulmonary disease berylliosis with symptoms of bronchitis, pneumonitis and interstitial alveolar oedema. Its chronic exposure damage alveolar membranes, inhibits phagocytosis leading to fine miliary nodulation in the lungs, fibrosis, granulomatous lesions and pneumoscerosis.

Epidemiological studies are difficult because only a small fraction of the exposed population develops symptoms of beryllium disease. A direct contact of skin with soluble beryllium compounds caused contact-dermatitis within a few days. The repeated exposure of the metal leads to allergic eczematous reactions. If a soluble compound is rubbed into a wound, a chronic ulcer develops, which does not heal until the offending material is removed.

Beryllium exposure to workers associated with atomic energy, ceramic and fluorescent lamp industries may cause bronchitis, pneumonitis, decreased macrophage, phagocytic capacity, alveolar interstitial oedema, fine miliary nodulation in the lungs, fibrosis, pneumosclerosis and lung cancer. Both acute and chronic beryllium intoxication have been reported among occupationally exposed subjects.

Experimental and epidemiological studies have shown that beryllium can cause cancer in lungs and bone marrow. It is not excreted from mammalian tissue and, therefore, its effects are cumulative. At the biochemical level, beryllium competes with magnesium for enzyme sites and has been shown to inhibit DNA polymerase, thymidine kinase and alkaline phosphatase. It selectively binds to regulatory non-histone nuclear proteins, inhibits phosphatase enzymes, interferes with gene transcription and inhibits DNA synthesis. The genotoxic effects were shown in various mammalian cell assay systems.

The absorption of beryllium from the gastro-intestinal tract is very poor (0.01%). Dermal absorption is also limited but inhalation constitutes a serious health hazard as substantial absorption may occur through this route. Experimental studies indicate that about 80% of the ingested metal passes into the faeces unassimilated and about 1% of the dose is excreted in urine. The metal is transported through the blood as colloidal phosphate or hydroxide complexes and may enter the cells by mechanisms involving cellular damage by lysozomal enzymes. It resides in the liver until excreted or translocated to bones. Retention is also reported in lungs and skeleton. Acute poisoning with soluble beryllium compounds is predominantly manifested by necrosis of the medium zone of the liver.

Bismuth

The Bulk of the orally administered bismuth remains unabsorbed and excreted in the faeces. Some mucosal uptake is also reported. Plasma concentrations peak within 15-60 min. of ingestion. The absorbed bismuth fraction is normally excreted in urine. Considerable renal accumulation has been observed to occur after its parental administration.

Cadmium

A highly toxic non-essential metal accumulates in the kidneys of mammals and can cause kidney dysfunction. In humans, kidney damage diagnosed by the presence of macroglobulin proteins in the main toxic effect resulting from chronic exposure to the metal. High concentrations of inhaled Cd aerosols can cause emphysema and related acute lung conditions. Cadmium becomes very volatile above 400°C and hence is likely to be dispersed as an aerosol when mixtures of metals containing Cd are heated or cast.

Cadmium is a normal constituent of soil and water at low concentrations. It is usually mined and extracted from zinc ores, especially zinc sulphide. Industrially, cadmium is used as an anti-friction agent, as a rust nuclear reactors electroplating bases, PVC manufacture and batteries.

It tends to be less strongly adsorbed than many other divalent metals and is therefore more labine in soils and sediments and more bioavailable. There is more danger from this metal moving through the human food chain from contaminated soils than most other metals. Sewage sludge-emended soils can contain sufficiently high concentrations of Cd to cause elevated concentrations of Cd in food crops and there is a European Community Directive limiting the maximum Cd content of sludged soils to 3 ppm. Although sewage sludge application are considered a major source of Cd in the soils receiving sludges, the most important sources overall are phosphate fertilizers and industrial emissions.

In the environment, cadmium is dangerous because many plants and animals absorb it efficiently and concentrate it within their tissues. Normally, however, retention from food by mammals is low but absorption is increased if the mammals are on a low calcium diet. Once adsorbed, cadmium associates with the low-molecular weight protein, metallothionein, and accumulates in the kidneys, liver and reproductive organs. Very small dose can causes hypertensions, heart and enlargement and premature death. There is some evidence suggesting that cadmium can induce chromosome abnormalities and may exert a carcinogenic effect on the lungs.

Gastro-intestinal absorption of cadmium occurs through passive diffusion and is very poor. Ingested cadmium is absorbed in small intestines (duodenum and ileum), deposited mainly in the liver, spleen and kidneys and excreted in urine (20%), bile and faces, (80%). Cadmium-uptake is related to body Ca-status, i.e. it is strongly inhibited by calcium-uptake. The pattern and extend of the deposition of inhaled cadmium depends on the particulate size (diameter 10-20 μm : nose and pharynx: diameter 5-10 mm tracheobronchial region: diameter 5μ m alveolar region). The metal is predominantly bound to erythrocytes in the blood. After absorption from the lungs and the gut, cadmium is transported via blood to liver and other body stores. It cross the blood-brain barrier and is retained in the brain, through in low quantities. The placenta and mammary glands act as barriers to the transport of cadmium to the foetus and the new born. Its concentration in the new-born is therefore extremely small.

Cadmium has a long biological half-life (>10 years) and its concentration in the body increases with age. The elements reacts with many macro-molecules and organic compounds of biological importance, e.g., purine, pyrimidines, nucleosides, nucleotides, RNA, DNA enzymes, etc. It competes with zinc to inhibit the SH-group of the thiol-containing enzymes. Cd-cysteinate, Cd-cysteinate $(OH)^-$, albumin -Cd complexes and aquo-Cd complexes all exists in the blood plasma. Albumin prevents the production of low molecular-weight fractions. There is convincing evidence to suggest that cadmium regulates the synthesis of mRNA's specific for metallothioneins (MT). Within the tissue, cadmium is mostly found bound to MT. There is a lag period of 3-4 h preceding the synthesis of MT, during which cadmium is distributed between the subcellular organelles and high molecular-weight cytosol proteins. Its biosynthesis occurs in response to cadmium as a protective detoxifying mechanism. As a result of such binding, the metal is rendered indiffusible and prevented from binding with proteins and enzymes.

Acute cadmium poisoning may result from drinking water or other fluids, particularly of acidic pH, contaminated with this metal such as resulting from the leaching of solders in water pipes, vending machines or cadmium-glazed pottery. Symptoms of exposure through ingestion or inhalation include nausea, vomiting, diarrhoea, short breath, headache, fever and chocking fits. Chronic effects are lung emphysema, renal dysfunction, inhibition of iron absorption, decreased haemoglobin levels, elevated blood pressure, disturbed liver function, increased excretion of calcium and phosphorus and bone demineralization. Intake a high-protein diet has a protective effect on cadmium toxicity. Protein malnutrition, on the other hand, exacerbates it adverse effects.

A serious case of Cd poisoning occurred in the Jintsu Valley in the Toyama Prefecture in Japan, where Pb-Zn mining had caused widespread Zn and Cd contamination on the alluvial soils, most of which were used for paddy rice production. There farmers in the valley live mainly on rice grown in the contaminated paddies and also relied on the metal-polluted river for their drinking water. After the Second World War, it was found that more than 200 elderly women who had all had several children had developed kidney damage and skeletal deformities. The condition was known as 'itai-itai' disease which literally means, 'ouch-ouch' due to the pain caused by the deformed bones. The Cd toxicity was exacerbated by a low protein and vitamin D diet and the birth of several children. The rice which they consumed contained tend times more Cd than local controls and the contaminated water was an additional intake. It was estimated that the poeple in the valley had a Cd intake of around 600 μg Cd/day which is around ten times greater than the maximum tolerable intake of 60—70 μg/day. A survey of paddy soils in the whole of Japan revealed that 9.5% of the area was significantly contaminated with Cd, with a further 3.2% of upland rice-growing soils and 7.5% of orchard soils. The source of the Cd in these soils is probably phosphate fertilizers and industrial/mining pollution.

Higher doses of Cd cause kidney problems, anaemia and bone marrow disorders. The major portion of Cd ingested into our body is trapped in the kidneys and eliminated. A small fraction is bound effectively by the body proteins, metallothionein which present in the kidneys, ; the rest is stored in the body and gradually accumulates. It excessive amounts of Cd^{2+} are ingested, it replaces Zn^{2+} at key enzymatic sites, causing metabolic disorders.

This occurrence of cadmium in rivers waters in different parts of the world is reported to be in the range 1 to 100 mg Cd per kg of the sample. Srivastava and Jaiswal (1989) have reported that whereas cadmium retards growth in the water plant *Spirodelapolyrrhiza* it induces formation of vegetatively propagating organ called *turion*. This is a distinct adaptive strategy to ensure successful recurrence of the species through turions after the stress is over in the water environment. They have noted the cadmium toxicity in this species is in the concentration range 0.005 ppm to 2.00 ppm and the adverse effects are on pigmentation and protein concentration.

The concentration of Cd in sea water on an average is 0.15 mg per litre (Goldberg, 1976). Anand (1978) has reported 16–17 mg Cd per kg of sample in fish sold in Bombay which is of the same order reported in other parts of the world. The toxic levels are 200 mg Cd per litre of water.

The exposure to cadmium in industries may occur from inhalation of cadmium dusts or fumes. Heated metal volatilizes and burns to form CdO fumes. Concentrations as low as 0.1 mg Cd/m^3 in air may cause toxic manifestations and levels of 1 mg Cd/m^3 and above may be fatal. The symptoms of Cd poisoning, often mistaken for "flu", include dry sore throat, cough, severe chest pain and marked dyspnoea. The pathological changes include interstitial pneumonitis and pulmonary oedema. Death following heavy exposure to cadmium is due to generalized lung damage. Rhinitis, dryness, and ulceration of nasal passages, loss of olfactory function, bronchitis, cough and emphysema may be observed in chronic cases.

Cadmium exposure has been associated with higher incidence of lung cancer. Oral uptake of the metal does not lead to tumour induction while its inhalation leads to carcinogenesis. About 1 µg Cd/m^3 in air inhaled corresponds to about 60 µg/g of the metal in lungs and leads to 50% incidence of pulmonary tumours. Smoking cigarettes significantly enhances the total body-burden of cadmium (upto 100 ng pe cigarette) with a proportionate increase in the risk of cancer.

Chromium

Like most of the metals, chromium is widely distributed. For most organisms, it is essential as a micronutrient in trace quantities for fat and carbohydrate metabolism. In industry, it is used in making steel alloys, in chromium plating and in leather tanning. Chromates are water soluble and can poison sewage treatment processes. The chromium ion can exist in four valency states; Cr^{2+}, Cr^{3+}, Cr^{5+} and Cr^{6+}. Of these, the hexavalent ion is the most toxic and it should be reduced to the trivalent state to form insoluble products before chromium waste in released into the environment. Soluble chromate concentrations of 0.5 μg/ml have been shown to be significant. Hence the redox conditions in the environment are very important, water-logged soils with reducing soils will have the less toxic Cr(III). However, in many freely drained aerated sols the predominant form is also Cr(III) because soil organic matter results in the reduction Cr(VI) to Cr(III). Soils developed on ultramafic rocks, such as serpentinites can contain very high concentration of Cr or geochemical origin and cannot be considered polluted but either 'naturally' or 'geochemically' enriched.

The absorption, retention and excretion of chromium depends on its ionic form. Inorganic Cr^{3+} is poorly absorbed. When linked to amino acids or other biomolecules, the element is more readily taken up by diffusion across the plasma membranes. Such organic complexes constitute the most efficient supply of chromium for man. Absorption of the trivalent form is 1% or less while Cr^{6+} is better absorbed. The latter form readily passes through the erythrocyte membrane and binds to haemoglobin. While the absorption of this metal in the elderly or in the maturityonset diabetics is not significantly different from that in normal individuals, two- to four-fold higher chromium absorption is reported in juvenile and insulin-requiring diabetes. The primary site in gastro-intestinal absorption appears to be the jejunum. Occupational exposure of workers might occur through inhalation or skin contact with Cr^{6+} in welding plants. Alveolar macrophages and epithelial lung fluid provide protection against its carcinogenic forms. The reduction of Cr^{6+} takes place in these cell types to produce Cr^{3+} extra-cellularly, thus preventing chromium from entering the lung cells is oxidized toxic forms. Diffusion through skin may also occur which depends on the nature of the compound.

The permeability of membranes to Cr^{6+} allows more chromium to enter and be reduced, resulting in a greater accumulation of the immunologically active chromium in the cell than that absorbed from direct contact with Cr^{3+} compounds.

After adsorption, Cr^{3+} binds itself to the β-globulin fractions of serum proteins, especially transferrin. Cr^{6+} accumulates inside erythrocytes where it is reduced to Cr^{3+}. This reducing ability of erythrocytes is an important mechanism is the detoxification of chromium. Transportation of chromium by blood to different organs and tissues occurs in protein-bound forms. Under normal conditions, transferrin is only 30% saturated with iron. The affinity of Cr^{3+} for transferrin is close to that of Fe^{3+} and, therefore, it competes with iron for the same binding sites. Hence, patients with haemochromatosis (with > 50% Fe saturation) retain less chromium than normal or iron-depleted subjects. A relatively higher concentration of chromium occurs in nuclear proteins. Within the tissues, chromium is associated with RNA in a non-dialiasable form. A chromium-containing nucleoprotein has been isolated from beef liver. Unlike most other trace elements, the tissue levels of chromium decline with age. Conversion of inorganic chromium to biologically active form is essential for its physiological function. Though the exact structure of such forms is not known, the biologically active form of chromium, isolated from brewer's yeast, reportedly contains nicotinic and acid and some amino acids such as glycine, glutamic acid and cysteine. The biological activity can be measured in chromium-deficient rats by (1) potentiation of the action of insulin in the glucose oxidation of adipose tissue and (ii) insolated epididymal fat cells.

The toxicity of chromium depends on the oxidation state of its compounds. Hexavalent chromium is more toxic than trivalent. Its accidental ingestion may cause acute human poisoning. Oral uptake of 0.5g of potassium dichromate is fatal for man. The metal is corrosive and may be absorbed through skin in dangerous amounts. The symptoms of its acute toxicity include diarrhoea, astro-intestinal bleeding, cramps and severe damage to liver and kidneys. The numerous industrial uses of chromium are the major causes of its exposure to many workers, leading to potential ill-effects, that include skin allergies, dermatitis, slow-healing ulcers, eczema, perforation of nose walls, dysomia, bronchial asthma, and lung cancer. A solitary study on a male patient reported an association of lepromatous leprosy and contact dermatitis due to exposure of chromium present in cement. Further studies to confirm or refute such links are not available.

Chromite(III) appears to be more toxic to fish than Cr(VI), especially, salmon, but toxic concentrations for several species of fish range from 0.2—5μg/ml. Municipal wastewater can contain concentrations of <0.7 μg/Cr ml, mainly in the Cr(VI) form, which are toxic to many species to marine animals, algae and microorganisms 'but reduction of Cr(VI) to Cr(III) usually occurs if there is organic matter present (Langard, 1980).

Dermatitis and eczemas caused by contact with cement have been linked to allergic reactions against Cr(VI) impurities in cement. In a Scandinavian study on a population group of over 5500 patient, 3% of the women and 12% of the men with skin allergy, showed, positive reaction tests with potassium dichromate. Epidemiological studies have also demonstrated an association between occupational exposure to this metal and lung cancer.

Occupational contact-dermatitis often occurs among workers of cement and tanning industries. Chromates may even cause chronic skin ulceration. A few reports also suggest an association between lepromatous leprosy and allergic contact-dermatitis due to the exposure of chromium in cement.

Workers exposed to the chromate dust the extended periods suffer from perforated nose walls, loss of smelling capacity and ulcers. Asthmatic bronchitis has been reported as an allergic reaction to Cr(VI) compounds. A sensitized person may have asthmatic attach 4-8 h following inhalation of chromate dust or even after dermal exposure to Cr(VI) impurities in cement. The industrial exposure through inhalation may also lead to lung cancer.

Chromium is carcinogenic, causing cancer of the respiratory organs in chromate workers chronically exposed to Cr-containing dusts (Langard 1980). Hexavalent chromium has been implicated in poisoning in Japan. In this case, aerosols from chromium refining plants appear to have affected a considerable number of people causing lung cancer. Besides this, it has been shown that chromates act as irritants to the eyes, nose and throat, and chronic exposure may lead to liver and kidney damage. A characteristic effect on human beings is the appearance of perforations in the nasal septum. At the cell level it appears that hexavalent chromium may cause chromosome abnormalities. Chromium is particularly dangerous because it accumulates in many organisms. Some aquatic algae have been shown to concentrate it 4000 times above the level of their immediate environment.

The maximum permissible limit of Cr in drinking water as recommended by the W.H.O., is 0.05 mg per litre. Rai and Raizada (1987 and 1988) have studied biochemically the effect of nickel, silver, chromium and lead on toxicity and its regulatory mechanisms on the nitrogen fixing blue green algae particularly *Nostoc muscorum* when treated singly and in different combinations.

Following oral ingestion or inhalation, chromium accumulates in the liver. In human beings, chromium intoxication leads to two types of syndromes: pulmonary, and gastric. However, in both the cases, it is associated with liver damage ranging from an increase in the size of the organ to cirrhosis and organ dysfunction. In an unexposed population, the blood-chromium level tends to decrease considerably in hepatic disorders. Experimental chromium deficiency in animals results in a reduced response of isolated liver mitochondria to insulin.

Cobalt

Most cobalt compounds have a valency of 2 or 3. The divalent form is more stable as simple salts while the trivalent form is stable in complexes. Normally, cobalt is ingested primarily in the form of various complexes and metalloenzymes. One such complex, synthesized in situ exclusively the microorganisms, is vitamin B_{12} also known as cyanocobalamine. It contains 4.35%.

The principal routes of entry into the body are inhalation, oral, skin and unprotected mucous membranes. After oral ingestion, a high absorption (60-90%) of cobalt from the gastro-intestinal tract high been recorded. Its occupational exposure occurs mainly through skin and inhalation. The uptake of cobalt is apparently coupled with that of iron. Anaemic patients responsive to the iron therapy exhibit increased cobalt adsorption. As cyanocobalamine, the metal contributes to the formation of red blood cells and is essential to the normal functioning of all cells, particularly those of bone marrow, nervous and gastro-intestinal systems. Cobalt inhibits cellular respiration of enzymes of the acid cycle. Such a hypoxic stimulation, mediated by cobalt, leads to an increased erythropoietin biosynthesis: which in turn, triggers the erythropoiesis in the bone marrow. Storage sites for this metal are bones, lives, spleen, kidneys, (the longest time of retention), endocrine glands and lymph nodes. Although inorganic (non-vitamin B_{12}) cobalt is invariably present in organs and body fluids, its physiological role is not known.

Cobalt may replace Zn, Mg and Mn in a variety of enzymes *in vitro* due to similarities in ionic size. Addition of cobalt *in vitro* has been reported to increase the activity of a variety of enzymes such as alkaline β-glycerophosphatase, dehydrogenase, hydrogenase, nitrate reductase, oxidase, pancreatic amylase and lipase, phosphatase, proteinase and zymosan. It activates the thyroid gland in patients with iodine deficiency and best results are obtained by the combined use of cobalt and iodine.

Excretion of ingested cobalt occurs mainly through faeces (60-80%) and urine (20-3-%). It is fast and intensive during the first 24 h and most of it is completed within a week. Beyond that, the metal is slowly excreted in micro-quantities in sweat, saliva, hair nails, and seminal fluid. Inhaled cobalt particles are rapidly cleared from lungs.

Its chronic toxicity may result from a combined therapy of anaemic patients with iron and cobalt. The symptoms include anorexia, gastro-intestinal disturbances, diarrhoea, vomiting, substernal aches, skin rashes, tinnitus, paraesthesia, neurogenic deafness, damage to the optic nerve, etc. The metal causes hypoxia through inhibition of enzymes involved in cellular respiration and citric acid cycle. This leads to enhanced erythropoietin biosynthesis and polycythemia. Industrial exposure through dusts of tungston carbide (commonly bonded with cobalt powder) and hard-metal grinding (using cobalt-containing coolants) may induce allergic alveolitis, eczema, increased erythropoiesis, decreases ability of thyroid to accumulate iodine, i.e., goitre, etc. Most of these effect are temporary and pass off following 1-2 months of cessation of exposure

Fatal cobal intoxication has been reported in heavy beer drinkers with symptoms of congestive heart failure, thyroid dysfunction, damage to the eighth cranial nerve, deafness and coma. This was attributed to excessive consumption (12 liters or more with cobalt intake of 6-8 mg per day) of certain brands of beer, to which cobalt sulphate (1 mg/litre) has been added to stabilize froth. It is interesting to note that intake of cobalt in much larger quantities in other forms (e.g. in anaemia therapy) was not lethal. These observation suggestion involvement of other factor in cobalt toxicity.

The occupation exposure to cobalt-containing dusts may lead to progressive pulmonary fibrosis and other forms of chronic lung damage. Such patients exhibit symptoms of dyspnoea and olfactory dysfunction and may also suffer from hard metal fibrosis of the lung.

Cobalt sulphide has been shown to be a more potent carcinogen than its oxide in implantation experiments in mice. There are reports of higher incidence of bronchial carcinoma among cobalt miners.

The liver is the major site for cobalt metabolism. Both as a free metal and in complex forms like cobalamin (vitamin B_{12}) or enzymes, it is accumulated in the liver. Although liver vitamin B_{12} is usually believed to constitute the body stores, experimental studies in rats indicate rapid fluxes of cobalt between liver and rest of the body. Cobalt-induced biosynthesis of erythropoetin occurs primarily in kidneys and liver. Its chloride salt has been used for experimental induction of nephrotoxicity in rabbits.

Copper

Copper is one of the most abundant trace metals. It is widely used in its metallic state, either in the pure form or in alloys. For almost all organisms it is an essential micronutrient. It may occur in very high concentrations in water, sediments and biota in some localized areas as a result of mining activities, of intensive use of copper pellets in pig rearing, or of the application of copper fungicides. However, there is no evidence of food chain magnification. Hence, most toxic effects are due to immediate exposure to the element. All organisms are harmed by excessive concentrations, which may be as low as 0.5 pm for algae. Most fish are killed by a few parts per million (Lopez and Lee, 1977). In higher animals brain damage is a characteristic feature of copper poisoning.

Copper is readily available as Cu (I) or Cu (II) in inorganic salts and organic complexes. About of 30-50% of the oral intake is absorbed mainly from the duodenal mucosa through a copper-binding protein. The extent of absorption depends on the form of copper, the acidity of gastro-intestinal tract and the presence of other substance in diet. Phytate, high ascorbic acid, certain amino acids and ligands which form complexes with copper, and metals, such as cadmium, mercury, molybdenum, silver and zinc reduce its absorption and alter its tissue distribution. Absorption may occur through skin. Upto 50 mg of the metal from copper bracelets (traditionally used to alleviate muscular and cardiovascular ailments) may be lost per month.

Absorbed copper is loosely associated with albumin and is distributed to almost all tissue and body fluids through plasma. Organic complexes permit transport through the cell wall. The metal forms an integral part of the DNA or RNA molecule. Ceruloplasmin in liver, and erythrocuperin in bone marrow are major copper-containing proteins. Its larger amounts are present in brain, liver, heart, kidneys and the pigmented parts of the eye. The key organ of its metabolism is silver where it is stored insoluble fraction (52%), nuclei (15%), mitochondria (12.4%), and microsomes (14%). The metal is a constituent of many enzymes such as hemocyanin, tyrosinase, lysal oxidase, ascorbic acid oxidase, cytochrome, oxidase, uricase, monoamine oxidase, dopamine hydroxylase etc. Copper is needed for the production of haemoglobin, oxygen transport and formation of aortic elastin and bone. Transfer or iron from tissues to the plasma requires the oxidation of the ferrous form. This is catalyzed by ceruloplasmin. The metal facilitates incorporation of iron into haemoglobin. The primary biochemical effect in bone disorders, neonatal ataxia and cardiovascular dysfunction is a substantial.

Unabsorbed copper, as well as a high proportion of injected copper, is excreted through bile and faeces. Urinary excretion is less than 50 μm per 24 h.

Copper can be deficient in some soil causing severe loss of yield in several crops, especially cereals. Toxicity problems can occur in crops in polluted soils and in livestock grazing herbage growing on polluted soils, sheep are the agricultural livestock most sensitive to Cu toxicity, but they (and cattle) are also prone to deficiency in both sheep and cattle but if the herbage contains > 10 ppm Cu then toxicity if likely to occur in sheep. Copper can arise from Cu mining and smelting, brass manufacture, electroplating and excessive use of Cu-based agrichemicals (e.g. Bordeaux Mixture). Copper sulphate is used widely as an algicide in ornamental ponds and even in water supply reservoirs which are affected by blooms of toxic blue-green algae. Copper is used widely in houses for piping water and although the concentrations of Cu in the Drinking water is higher in soft water, this is not considered to be a hazard so long as the pH i within the normal limits (pH 6.5—8.5). More acid waters could create problems with excessive concentrations, but none have been reported with public water supplies.

Excessive copper accumulation is observed in kidneys of patients suffering from Wilson's disease. This accompanied by low serum- ceruloplasmin levels and high urinary excretion. The latter may reach 1500 μg day as against mean urinary copper excretion of 215.2 μg/day in normal subjects. In renal failure/uremia, serum copper levels are significantly raised.

The inhalation of dusts, fumes or mists of copper compounds causes congestion of nasal and mucous membranes with occasional perforation of the nasal septum. Its chronic exposure may lead to pulmonary copper deposition and fibrosis.

Abnormally high serum-copper levels are observed in chronic infectious diseases (e.g. tuberculosis) and bronchial carcinoma.

Links between contamination of drinking water by copper due to boiling/storage in brass/copper utensils and Indian Childhood Cirrhosis (ICC) have been reported by many investigators.

Copper salts have also been associated with the 'vineyard sprayer's disease,' which affects both lung and liver. It is observed among grape garden sprayers using the bordeaux mixture (which contains copper sulphate and lime). Its prolonged exposure causes hepatosplenomegaly, granulomas and portal stenosis. Periodic estimates of the serum alkaline phosphatase level, which is usually elevated have been recommended for identifying workers prone to hepatic damage from such an occupational exposure.

Copper nas a special role in vitiligo or leucoderma. Though the precise etiology is not fully understood, the disease is attributed to disturbances in the synthesis of melanin, which involves tyrosinase, a Cu-containing enzyme. Both elevation and depletion in serum-copper levels in patients suffering from leucoderma have been observed.

Gold

The main source of gold exposure to humans are gold containing drugs. Approximately 25% phosphine containing Au (I) complexes, e.g., auranofin, which are used as anti-arthritic drugs, are absorbed from the gastro-intestinal tract, reaching peak serum levels within hours. In the blood, the thiolate forms of gold react with albumin and other serum proteins involving ligand exchange reactions. The bulk of intravasal gold is protein bound of which more than 85% is bound to ablbumin. Transport across biological membrane has been observed and the metal may enter the ertythrocytes and other cells.

Gold concentrates in the liver, kidney and the synovial fluid. The metal may persist in the body for prolonged period. The principal depository is lysosomes which, upon accumulating the metal, become electron-dense and are called aurosomes. Retention in autopsy samples is discernible even after 20 years of cessation of gold therapy. Detoxification mechanisms include reduction of Au(III) to Au(I), possibly by protein disulfides, thiols and thioethers. Interaction of myocrisin with CN have been reported.

Gold is mainly excreted through urinary and faecal routes. For orally administered preparations such as auranofin, less than 20% of the clearance is renal, in contrast to over 60% for parenteral preparations of myrocrisin or solagnol.

Gold toxicity may occur as a side-effect of gold therapy in patients of rheumatoid arthritis with symptoms of mucocutaneous reactions, skin rash, oral ulceration, renal and hepatic damage, peripheral neuritis, encephalitis with EEG abnormalities, suppression of cells in the bone-marrow (red and white blood cells, and platelets) thrombocytopenia, and aplastic anaemic, which may sometimes be fatal. Skin reactions to gold jewellery, though rare, may be observed among some chrysotherapy patients.

Gold is not carcinogenic. Certain gold compound exhibit antipemphigus, cytotoxic and anti-tumour effects. Radioactive gold (Au) has been used in the form of a colloidal solution for radiotherapy in cancers of the gastro-intestinal tract, skin connective tissues, bladder, pleura, ovaries and prostate and pituitary glands. Such thereby, however, has limited utility and manifests many side-effects.

Iron

Iron is the fourth most abundant element in the earth's crust. Its greatest use is structural iron and steel, but it is also used for making dyes and abrasives. It is an essential micronutrient required in trace quantities for the normal metabolism of plants and animals. It is a constituent of cytochromes and nonheme iron proteins involved in photosynthesis, n2 fixation and respiratory linked dehydrogenases (Noggle and Fritz, 1986) Ingestion of excessive amounts may result in the inhibition of activity of many enzymes. The amounts consumed must be very large because only a small proportion of all iron igested is absorbed from the gastro-intestinal tract. Inhalation of iron dust can cause benign pneumoconiosis and can enhance harmful effects of sulphur-dioxide and various carcinogens.

Many streams are poisoned by high levels of iron in acid mine drainage. Pyritic, iron sulphide, is often found in close association with coal deposits. Upon exposure to moisture and atmospheric oxygen, the ferrous iron is oxidized to the ferric state, a reaction which is frequently accelerated by bacteria of the Thiobacillus—Ferrobacillus group. The ferric iron can then react with sulphide in the presence of water to produce sulphuric acid, or react directly with water to produce a yellow, flocculent mass of ferric hydroxide. Beside being acidic, water affected in this way becomes deficient in oxygen. Such poisoning of streams is reckoned to be one of the main causes of fish kill. Although particularly associated with mining, streams running through iron-laden strata may become poisoned spontaneously.

Acute iron toxicity may be the result of an overdose of medicinal iron. Its symptoms include haemorrhagic gastritis intravascular clotting, acidosis and fatal shock. Iron over-load has been associated with haemochromatosis, a genetic disorder of iron metabolism characterized by a brown discoloration of the skin and other organs due the destruction of red blood cells and deposition of iron pigments. The deposition of large amounts of iron in hepatic parenchyma causes tissue damage by the enhanced generation of the hydroxyl radical. Excess iron often prediposes to infection. A skin disease called Porphyria cutanea tarda, which is characterized by photosensitivity, blistering, scarring pigmentation and hyperkeratosis, has also been attributed to the overload of iron.

Shortness of breath is a common symptom of iron deficiency. Excess iron catalyzes the production of $^{\bullet}OH$ and O_2^- radicals. These radicals may induce peroxidation of lysosomal membranes, deploymerize hyaluronate, release serotonin and cause endothelial and lung damage.

Iron is suspected to be carcinogenic. Fats heated upto 300°C in iron vessels or in glass vessels in the presence of iron fillings also acquire this property. Painting the carcinogen on the skin of experimental animals leads to loss in epidermal iron within 10 days (upto 50%) and its continued application causes a further loss. In liver tumours, a considerable reduction in iron-containing enzymes such as cyto-chrome P-450 and catalase is noted. Self-medication with iron-containing tonics by middle-aged persons has been reported to cause a number of deaths from intestinal and stomach cancer. An enhanced mortality due to increased lung cancer has been reported following occupational exposure to iron dusts. Besides, iron particles probably serve as carriers for other carcinogens. It has also been observed that a population with iron overload is at higher risk of cancer incidence. The risk is further aggravated in smokers.

Lead

A non-essential element; it is a neurotoxin and a good example of a multimedia pollutant. The main sources of Pb pollution in the environment are petrol (air pollutant, but can also be water or soil pollutant from spillages), particulates in exhaust fumes from petrol combustion (air pollutants, inhaled by humans), particulates from petrol, fossil fuel combustion in soil (soil pollutant—taken up by plants and also ingested with plant food crops), paints flakes from old paint containing a percentage of Pb, pB in some traditional ethnic cosmetic (e.g. surma—skin absorption), constituent of solders and varnishes used on interiors of food can (food contaminant), pB pipes for portable water (water pollutant), pesticide (Historic use of pB and As containing pesticide sprays in orchards), had shot used in guns for game and clay pigeon shooting (soil pollutants, but also a food contaminant if inadvertently consumed with the game flesh) and, finally, pB pollution from mining and smelting of the ore (usually PbS)—this includes acid mine drainage with soluble Pb (water pollutant), tailing from ore dressing (particular water pollutant and soil pollutant, weather to release soluble Pb), and smelter fumes—aerosols (air and soil pollutants).

On a comparative basis, Pb is neither as toxic as many other metals nor as bioavailable, however, it is generally more ubiquitous in the environment and is a cumulative toxin in the mammalian body, so toxic concentrations can accumulate in the bone marrow, where red blood corpuscle formation (hematopoieses) occurs. At least five stages in the formation of the haem part of haemoglobin are affected by Pb but the two enzymes most affected are δ-amino laevulinic dehydratase (ALAD) and ferrochetalase (Waldron, 1980). This inhibition of haem synthesis results in anaemia. Kidney damage also occurs as a result of exposure to Pb. Lead, like Hg, is a powerful neurotoxin and range of pathological conditions are associated with acute Pb poisoning, most characteristics of which is cerebral oedema. However, the absorption of Pb in amounts which are not high enough to cause acute poisoning may induce behavioural abnormalities, including learning difficulties.

Industrialization and increased transport activities have led to lead pollution of the natural environment on a global scale. The metal has been shown to impair neurological, cardiovascular and reproductive functions and cause carcinogenic effects. According to a UNEP/WHO report, vast variations are found in the blood-lead levels among the inhabitants of big cities of the world, e.g., Tokyo (60 μg/1) and Mexico City (225 μg/1). This is alarming since at blood levels > 80 μg/dl, there is high risk of adverse cardiological effects.

Several studies report increased mortality among lead smelters and battery plant workers from gastro-intestinal, respiratory and renal cancers. Exposure to lead for more than 20 years may lead to increased incidence of cancer.

Absorption of ingested lead from the gastro-intestinal tract is very poor (5-15%) : about 293 mg of the average daily intake of 300 mg is excreted. The rest (mean : 7 mg, range :3-211mg/day) is mainly retained in the bones. Absorption is affected by a number of factors *chemical form* and solubility and acetate, chloride, oxide and tetraethyl compounds are readily absorbed whereas sulphides, sulphates chromates and carbonates are sparingly soluble but fairly absorbable. The rate of absorption is significantly enhanced on an empty stomach low calcium and vitamin D, iron deficiency and presence of ascorbic acid, and amino acids with SH-groups and bile acids. Elements competing for the same carrier systems in the intestinal epithelium, such as calcium, iron, magnesium, phosphate, ethanol and high-fat diets reduce the absorption of lead. Young children absorb more lead than do adult humans. Pulmonary absorption of inhaled lead depends on the state of aggregation (gas or solid), particulate size, concentration in the ambient air respiratory volume and distribution of lead within the respiratory volume and distribution of lead within the respiratory tract. Larger particles are cleared from the upper respiratory tract by ciliated epithelial cells. A portion of the lead retained in alveoli may, however, be cleared by macrophages.

Tracer studies indicate distribution of head in the body as follows :

i) blood with very short half-life (hours/days),

ii) soft tissues with medium half-life (weeks), and

iii) skeleton representing a poor very half-life (months/years).

More than 90% of the blood-lead is bound to haemoglobin in erythrocytes. A specific lead-binding protein has been identified and attributed with protective function. The metal in blood is largely exchangeable with tissues/organs. Though the principal binding sites for lead are macromolecules, mainly proteins with several free sulphydryl groups, e.g. a metallothionein containing cysteine), it may form less stable compounds with other side chains of amino acids, e.g. E-amino groups of lysine, carboxyl groups of glutamic and aspartic acids and phenoxyl group of tyrosine. Interactions with cadmium, calcium, copper, iron, magnesium, phosphorus, zinc, vitamin D and several enzymes, particularly in the heme pathway, have been reported.

Lead has been reported to cross the blood-brain and placental barriers. Within the brain, the highest lead concentrations have been found in the cortical grey matter and basal ganglia. Children are particularly prone to lead intoxication. A significant correlation exists between blood-lead concentrations of mother and children. Foetal uptake of lead is apparent by the 22nd week of gestation and increased throughout development. Excretion of unabsorbed lead occurs through faeces. (this being the main route of its elimination from the body) Sizable excretion may also occur through the sweat.

Lead compounds are reported to be mutagenic and genotoxic, causing disturbance in enzymatic functions and inducing chromosomal aberrations. Until early 1960s, most lead studies gave negative results regarding its carcinogenicity. However, it was shown later that acetate, sub-acetate, and phosphate of lead, when injected or administered through diet, could induce renal tumours in rodents. Kidney tumours develop at very high doses of lead and only in animals with obvious symptoms of acute lead toxicity with the appearance of nuclear inclusion bodies in renal cells. Some human studies reveal an increased incidence of cancer-mortality (gastrointestinal tract, respiratory and renal cancers) among lead smelter and battery plant workers. However, only long-term exposures have been linked to the induction of cancer.

The critical concentrations for Pb in blood are the EC recommended level of 35 g/dl and the UK threshold level for follow-up investigations of 25 mg/dl in at least half the population and less than 30 mg/dl in 90% of the population. The critical level in blood for occupational exposure in higher with values of 60, 80 and even 100 g/dl being used in different countries.

Manganese

Manganese occurs widely in nature and is of considerable importance in the manufacture of steel. Biologically, it is an essential micronutrient for most organisms. It is required for activity of some dehydrogenases, decarboxylases, kinases, oxidases, peroxidases, etc. It is required for photosynthetic evolution of oxygen. This metal is not a serious pollutant as in most waters its concentrations is quite low ranging from 0.005 to 1 mg per litre. Potassium permanganate is used in very small doses to disinfect well water particularly in Indian villages. Excessive amounts of manganese effects animals adversely, causing cramps, tremors and hallucinations, manganic pneumonia and renal degeneration.

The absorption of manganese occurs through respiratory and gastro-intestinal tracts. Organo-manganese compounds may also be absorbed through the skin. A portion of inhaled manganese is cleared by mucociliatory actions. The mechanisms of its absorption are not precisely known but it occurs throughout the length of small intestine. Its absorption rate is considerably increased with alcoholism, in iron deficiency anaemia (upto 7%), and during pregnancy and infancy.

Manganese toxicity is a serious threat to health in some industries. The new-born and infants constitute high-risk groups for manganese intoxication because of their low capacity to excrete the metal. The clinical features of its toxicity include broncho and lobar pneumonia not responsive to antibiotic therapy, akinesia, rigidity, increased muscle tone, altered, gait, trembling hands, emotional instability, insomnia, hyper-irritability, hallucination, mania, violent acts, compulsive phenomena and frank psychosis. These symptoms resemble those of Parkinson's disease. The mental symptoms recover on cessation of its exposure but extrapyramidal abnormalities are usually permanent. Another disease called sclerosis amyotrophic lateralis, which is reported from Japan, has been linked to the high manganese content in soil and food chain.

The half-time of manganese absorbed orally is about 10 days and that for parenterally injected about 40 days. The absorbed manganese enters the portal blood where most of it gets bound to x-macroglobulin. It is also incorporated in erythrocytes during erythropoiesis. In the liver, a fraction of the element is oxidized to Mn by ferroxidase I and then bound to transferrin. The organs where manganese is mostly concentrated are liver, pancreas, kidney and the intestines. Within the cells, manganese is found in mitochondria. The metal acts as a cofactors in several metaloenzymes, such as arginase, pyruvate carboxylase, superoxide dismutase, and elicits important functions in the brain, in the formation of connective tissues and bones, and in the metabolism of biogenic amines, carbohydrates and lipids. Manganese catalyses the formation of glucosamine-serine linkages during the synthesis of mucopolysaccharides of cartilage. The element can penetrate both blood-brain and placental barriers. The half-time for brain is longer than that for the rest of the body.

The excretion of inorganic manganese occurs mainly through

the bile and faeces. Its marginal excretion occurs through pancreatic juice and kidney. Only 0.1–1.3% of the daily intake is excreted through urine. However, renal excretion for organo-manganese tricarbonyl compounds, which are used as additives in gasoline, is substantial.

Studies in micro-organisms give some indication regarding the mutagenic actions of manganese. Its mutagenic or carcinogenic effects have not been confirmed in mammals. Some reports, on the contrary, show higher death rates from cancer in areas where its content in drinking water is low.

Mercury

Mercury is concentrated in various ores, the principal one being cinnabar (HgS). It has been mined since 700 BC and is currently used industrially in three forms : as the metal, in organic compounds and in inorganic compounds. The greatest use of mercury is in the production of electrical apparatus. The second greatest use in the chloro-alkali industry, which produces chlorine and caustic soda by electrolysis of sodium chloride solution using mercury as the cathode of the electrolysis cell. The third greatest use of worldwide is in fungicides.

Nearly all the mercury used by man eventually enters the natural environment. To this may be added amounts released during the production of mercury and other metals, from coal burning and from weathering of rocks. All forms of mercury are potentially toxic but the toxicities vary considerably. The least toxic are the inorganic mercury compounds. They are not readily absorbed from the gastrointestinal tract. Once absorbed they may accumulate in the liver and kidney but normally they are excreted quite rapidly in the urine. It is worth noting that mercury amalgam has long been used in density to fill teeth without any toxic effects being noted.

Hg^{2+} ion, is fairly toxic. It has high affinity for sulphur atoms, and easily attaches itself to the sulphur containing amino acids of proteins. It also forms bonds with haemoglobin and serum albumin, both of which contain sullphydryl groups. This ion, however, does not travel across membranes and hence does not get access to cells.

The main route of entrance of mercury into the body is through inhalation, as expected from its monoatomic nature, high vapour pressure, lipid solubility and high diffusibility. Approximately 80% of the inhaled mercury is retained. The absorption of the inorganic mercury from food is about 7%, the absorption of methyl mercury from diet is about 95% in adults. Most forms of mercury can penetrate the skin to some extent; systemic absorption of alkyl mercurials is substantial and poisoning may result from dermal application of such compounds. The metals is disturbed to almost every tissues or organ of mammals, including brain, placenta and foetus. However, its highest concentrations are found in the kidney of an exposed animal. Mercury vapours are rapidly oxidized in tissues to divalent ionic form. Organo-mercurials such as phenylmercury and mercurial diuretics also undergo rapid transformation into inorganic mercury. Short chain alkyl-mercury compounds (methylmercury) undergo slow biotransformation; The toxic properties of methylmercury may partly be attributed to the stability of the C-Hg bond. The ratio of red cells to plasma content depends on the form of mercury; it is highest for methyl mercury and lowest for inorganic mercury. Hair content is a useful index for assessing blood levels of methyl mercury; the concentration in hair being about 250 times higher than the concentration in blood.

Mercuric ions , Hg^{2+} have a high affinity for sulphydryl (SH) groups. As many protein contain SH-groups or disulphide brides, the mental can cause widespread alteration in their functions. The distribution in the brain and the central nervous system is largely governed by the available mercury binding sites/SH-groups.

Mercury vapour is the most hazardous of the inorganic forms because it can diffuse through the lung into the blood and then into the brain, where serious damage can occur. Arylmercurials are as toxic as the inorganic forms since they are readily broken down to inorganic derivatives in the tissues. Alkylmercurials RHg^{+} are the most toxic mercury compounds so far studied. They are fairly stable and have long retention times in the tissues. Therefore, they readily accumulate to high concentrations. There lipid solubility gives them an affinity for nervous tissues which account for many of their harmful effects. In addition, they have been reported to the cause abnormalities in cell division and increase the frequency of chromosome breakages. Some of the abnormalities may be due to combination of mercurials with sulfhydryls groups. Inhibition of enzymes in this way has been demonstrated frequently. So far, no effective treatment for mercury poisoning has been developed. Though British antilewisite (BAL) or calcium ethylenediaminetetraacetate (Ca_2EDTA) may have some alleviating effect.

Symptoms of mercury poisoning include gingivitis, loosening of teeth, metallic taste, stomatitis, gastroenteritis, dysphagia, diarrhoea, cough, chest pain, dyspnoea, paralysis of the muscles of larynx and pharynx, bronchitis, pneumonitis, renal damage (polyuria followed by oligouria and anuria), arrhythmias, cardiovascular collapse, neuropsychiatric manifestations, skin hypersensitivity reactions, immunotoxic, lmutagenic, carcinogenic and teratogenic effects. Critical urinary concentrations above which toxic manifestations may be visualized range between 1-2 μg/ml. Oral ingestion of one gram of mercuric salt is lethal.

Mercury intoxication, by both inorganic salts and organic compounds, may lead to neurological damage and grave genetic defects. Alkyl derivatives are more toxic than other forms. The damage caused by such compounds to the nervous system is often irreversible. The Minamata disease is a classical example of poisoning with the methyl mercury on an epidemic scale. The first case reported by Dr. Hosokawa in 1956 near Minamata bay in southern Japan. It was shown to be caused by eating fish contaminated with CH_3HgCl discharged by a chemical plant into the bay. There mercury load was so great that it killed fish and fish-eating birds, which dropped into the sea with intense symptoms of violent whirl, dashing with a zig-zig course and collapse. Though links between water pollution, consumption of contaminated fish and human disease were established in 1959, the effluent dumpting could be halted nine years later in 1968.

It is manifested by fatigue, loss of memory and concentration, headache, insomnia, drowsiness, numbness in limbs and face, paraesthesia, ataxia, tremous (fine trembling interrupted every few minutes by coarse shaking), dysarthria, constriction of visual field, cortical blindness, hearing defects, acute psychosis, slurred speech, imprecise handwriting with missing words, uncordinated gait, etc. A typical comatose posture with arched neck, tightly flexed arms, clenched fists and crossed legs have been described. Methyl mercury-induced behavioural effects are reversible but damage to the foetal nervous system is usually permanent. Principal damage is caused to granular layer in cerebellum and the visual cortex with neuronal loss and glial proliferation. The biochemical site of action is the sulfhydryl group of enzymes.

Cases of sudden death have been reported following i.v. use of mercurial diuretics in congestive heart failure. Death occurs within few minutes of injection. It is believed to be due to ventricular arrhythmias. Cardiovascular collapse is more common in patients with nephrotic syndrome. Atreoventricular dissociation, extrasystoles and ventricular tachycardia precede fibrillation.

The recognized supremacy of non-mercurial preparations has largely replaced mercurials for antiseptic, parasiticidal and fungicidal effects. The alkyl and alkoxy compounds are very toxic, cause local irritation and systemic poisoning when absorbed in the body. Hypersensitivity skin reactions may result from the use of mercurial diuretics in some patients. These may vary from erythema involving the area of contact to severe but generalized eruptions. A mecury-induced syndrome, known as the pink disease, has been reported in children. It is characterized by erythematous fingers, toes, cheeks, nose and buttocks.

Consumption of bread contaminated with mercury-containing fungicides led to intoxication in over 60,000 people, more than 6000 hospital admissions and 200 deaths in Iraq in 1971–72. The symptoms resembled those of the Minamata disease. Similar epidemics were reported from the Nigata region of Japan, Canada, USA and some other countries.

Nutritional contamination with organic mercury is highly neurotoxic, specially for young children and pre-netal development of foetus. Children born to mildly diseased mothers (with or without symptoms) suffered from cerebral palsy and idiocy. Women with more serious disease showed higher incidence of abortions and still births. Epidemiological studies have also shown links between mercury exposure and leukemias. Selenium elicits a protective role against mercury intoxication. More research is needed to probe these aspects.

Inorganic and organic mercury compounds elicit antiseptic, parasiticidal and fungicidal properties. The alkyl and alkoxy compounds are very toxic, causing local irritation and systemic poisoning when absorbed. Hypersensitivity to skin reactions may result from the use of mercurial diuretics in some patients. The symptoms may vary from mild erythema involving the area of contact to severe generalized skin eruptions.

The excretion of mercury follows several routes. It can be volatilized from the lungs and skin, can be eliminated via urine, sweat, milk, saliva, intestinal mucosa, bile, faeces, or stored in and shed through hair and nails. The average half-life of methyl mercury in man is about 70 days (range 46-120 days). The principal route of excretion is through faeces, but in cases of prolonged exposure, urinary excretion may marginally exceed faecal elimination. Concentration of total mercury in urine has no correlation with blood mercury levels in populations heavily exposed to methyl mercury.

Many ayurvedic and unani drugs used in India contain high concentrations of mercury. Enzyme inhibitory and hepatoxic effects with some of such preparations have been reported. In the Minamata disease, liver accumulates substantial amounts of mercury.

Biological Methylation

Unfortunately, in assessing the risk mercury in a particular environment, it is not enough to know the form in which it entered that environment because various transformation can take place. Probably the most serious of these is the transformation of metallic mercury to methyl and dimethyl derivatives by anaerobic micro-organisms, especially *Clostridium cochleariul*, in aquatic sediments. This may also occur in decaying fish the transformation is facilitated by Co (III)- containing vitamin B_{12} coenzyme. A CH_3– group bonded to Co (III) on the coenzyme is transferred enzymatically by methyl cobalamin to HG^{2+}, yielding CH_3GH^+ or $(CH_3)_2$ Hg. Under aerobic conditions, this transformation can be brought about by *Pseudomonas* spp, and by the fungus, *Neurospora crassa*. In essence, it represents conversion of the least toxic form of mercury to the most toxic. Other transformations which can be brought about by bacteria are the following :

Phenyl, ethyl and methyl can be reduced to elemental mercury and benzene, ethane and methane, respectively.

Phenylmercuricacetate can be converted aerobically to elemental mercury and diphenyl mercury;

Mercuric ions can be reduced to elemental mercury. Not all the transformation observed are biological. Under alkaline conditions, methyl mercury converts to the more volatile dimethyl mercury. Under oxidizing conditions, in the presence of ultraviolet light, phenyl mercury, alkoxyalkyl mercury and alkyl mercury may break down to give inorganic mercury. Under anaerobic conditions, mercuric ions may combine with hydrogen sulphide to form poorly-soluble mercuric sulphide. With subsequent aeration, mercuric sulphide can be converted to the soluble sulphate which may then be methylated biologically.

Most macro-organisms are relatively insensitive to mercury and its derivatives. Nitrogen fixing bacteria in the soil require levels of about 100 ppm before they are adversely affected. Normal soil levels are between 0.005 and 1 ppm. However, marine and freshwater phytoplankton, especially diatoms, are very sensitive to organomercurial fungicides and as little as 0.001 ppm may reduce their photosynthetic efficiency. Many algae and other plants have the ability to absorb and concentrate mercury from the surrounding environment. Droplets of elements mercury have been found in chickweed. Such high concentrations may cause mitotic disturbances and kill the plants. Fortunately, most agricultural plants do not seem to absorb much mercury. Animals tend to accumulate mercury through there food. Pike* can accumulate a concentration of mercury 3000 times higher than that in the water in which they live. Tuna and Swordfish show the same ability. Similar observations have been made on predatory birds. Seed-eating birds accumulate mercury where alkylmercury seed dressing are being used. Much less is accumulated where alkoxyalkyl compounds are used, and negligible quantities where inorganic mercury compounds are used.

The effects of mercury and it derivatives on man deserve special consideration because it was mainly these that caused concern about the effects of heavy metals released into the environment in large amounts. The first serious incident to come to light occurred at Minamata Bay in Japan. In this case, comparatively non-toxic inorganic mercury along with some methyl mercury was released in effluent by a chemical factory using mercuric sulphate catalysts in acetaldehyde production. The effluent entered river running to methyl mercury. The accumulated in shellfish and fish which were eaten by the local inhabitant. Consequently by 1975, 116 people

---

* A pike is a large fish that lives in rivers and lakes, and that eats other fish.

had died and many were left paralysed for life. Others suffered impairment of vision and hearing and other neurological symptoms. prenatal poisoning of the foetus was observed even in the absence of symptoms in the mother. Since the Minamata Bay incident, another has occurred around the Agana River, Nigata, Japan. This led to 23 deaths. In both these cases, many domestic animals, fish, shellfish and seabirds were affected.

Mercury poisoning of human beings has also been caused by the consumption of food containing high concentration derived from alkylmercury agricultural seed treatments used to prevent seed-borne disease. In Iraq, treated seeds, intended for planting, were used to make bread. Thousands were poisoned and hundreds died. In the USA, cattle fed on treated grain were slaughtered from human consumption. Again, severe poisoning resulted. Consequently a number of countries have now banned the use of alkylmercurial seed treatments. A committee of experts constituted by the Food and Agriculture Organization (FAO) and World Health Organization (WHO) has recommended that the use of alkylmercurials should be restricted to stock of cereal seed used for plant breeding or seed production, and never permitted for treatment of cereal seed for export for the production of food.

Despite the major incidents referred to above, it seems fairly certain that the average person is at no great risk from exposure to mercury. The normal dietary intake a well below what is thought to be the tolerable limit of 0.3 mg per person per week, of which not more than 0.2 mg should be in the methylated form, according to WHO. For most people, the chances of appreciable exposure from air, pesticides or pharmaceuticals are very limited. Where there is a risk of occupational exposure, this should be minimized by appropriate, there is still need to know more about the concentration and distribution of mercury, especially with regard to those areas where localized high concentrations do exist.

Nickel

Nickel is used in various forms for nickel plating, as a catalyst, as a mordant and in ceramic glazes etc. Again it is a micronutrient for most organisms but excessive quantities have toxic effects. In animals these include dermatitis and respiratory disorders, including lung cancer following inhalation. Amongst enzymes inhibited are cytochrome oxidase, isocitrate dehydrogenase and maleic dehydrogenase. A particularly poisonous derivative of nickel is nickel tetracarbonyl, $NI(CO)_4$.

Less than 10% of the ingested nickel is normally absorbed. A higher percentage may be absorbed during pregnancy. Food and beverages reduce or prevent the absorption of Ni^{2+} from the gut. In humans, approximately 35% of the inhaled nickel is absorbed; the remaining part is cleared out by the mucociliatory action in respiratory passages and their swallowed or expectorated. Ni-carbonyl, is highly volatile and is absorbed readily from lungs.

Transport of Ni^{2+} in plasma is mediated by binding to albumin and unfiltrable ligands. The latter include amino acids (e.g. the histidine and cystine) and small polypeptides. Nickel carbonyl is lipid soluble and penetrates the blood-brain barrier, but nickel concentration in brain has been reported to be substantially lower than that in other tissues. Within the body, it is widely distributed in most tissues and organs. The metal has been consistently reported to be present in RNA. It is a potent activator of several enzymes and probably plays and role as a bioligant in iron absorption, regulation of prolactin, and in the structure and function of membranes. Interactions with calcium, chromium, copper, cobalt, iodine, iron, magnesium, manganese, molybdenum, phosphorus, potassium, sodium and zinc have been reported. Placental transport and appreciable retention in foetal tissue may occur.

Most of the ingested nickel remains unabsorbed and is excreted in faeces. Urinary excretion is small and in the form of low molecular complexes which include the complexes with histidine and aspartic acid. Healthy adults contain significant amounts of nickel in sweat, suggesting its active secretion of sweat glands. Excretion also occurs through milk, hair and exhaled air.

Exposure to nickel through dermal, oral or inhalatory routes has been reported to cause allergy, dermatitis, eczema, rhinitis, sinusitis, bronchitis, pneumonitis, asthma, perforation of the nasal septum, inhibition of phagocytosis and the T-cell-mediated immune response, and mutagenic, carcinogenic and teratogenic effect in animals and man. The most common form of human toxicity of nickel is dermatitis. Epidemiological studies showed high incidence of lung, larynx, and sinus cancers among workers exposed to nickel in refinery, smelting and electrolysis works.

Lung is the primary target organ for nickel toxicity. The signs and symptoms include dyspnoea, cough, intense chest pain, cyanosis and bronchoalveolar hyperplasia. Industrial exposure may cause lung cancer. Employees of electroplating and electrolytic refinery works, who inhale vapours of water soluble nickel salts, develop chronic respiratory diseases, such as asthma, bronchitis, rhinitis, sinusitis, nasal polyps, pneumoconiasis and perforation of the nasal septum.

Allergic contact dermatitis has been reported in sensitive female populations from hair-pins, imitation jewellery, brassier hooks and other metal fastenings on clothing. Dermatitis is also seen in the palms of hair dresses from frequent use of nickel-containing shampoos. Nickel sulphate solution (5%) is included in the European battery of standard patch tests. It is interesting to note that experimental nickel deficiency in chicken, rats and swines cause skin discolouration, dermatitis and loss of hair. These symptoms are corrected by nickel supplementation in diet.

Experimental and epidemiological evidence indicates nickel to be mutagenic and carcinogenic in both animals models and man. Administration of soluble salts in drinking water promotes the carcinogenic action of nitrosamine in rats. The metals has been related to the incidence of malignant refinery workers exposed to its dust and nickel-carbonyl.

Selenium

Strictly speaking selenium is not a metal, though it has certain metallic properties. It is member of the sulphur-group, produced as a by-product of the extraction of copper, nickel, gold and silver ores. It is used in electronics, and in paints and rubber compounds, for most organisms it is an essential micronutrient but it can be toxic very low concentrations. The maximum permissible concentration in drinking water is 0.01 ppm. Poisoning of livestock has occurred where cattle have eaten plants of the brassica family which have taken in selenium and incorporated it is cysteine and methionine in place of sulphur. Selenium itself irritates the eyes, nose, throat and respiratory tract. It can cause cancer of the liver, pneumonia, liver and kidney degeneration and gastro-intestinal disturbance.

Experimental toxicity in rats, hamsters and rabbits causes anorexia, enteritis, growth depression, weight-loss, renal and hepatic damage, anaemia, postural defects, paralysis of hind-quarters, reduced fertility and teratogenic effect. Ingestion of selenoamino acids in plants (3-20 ppm SE) has led to acute and chronic toxicity in farm animals. Two distinct diseases related to excessive selenium exposure have been described : (1) blind staggers with symptoms of hypersalivation, garlic-like odour in breath, emesis, diarrhoea, dyspnoea, pulmonaıy oedema, tachycardia, depression, ataxia, incoordination, paralysis and death, and (ii) alkali disease with predominant hepatotoxic effects, growth depression, low haemoglobin levels, ascites, and atrophy of liver and heart. The internal organs show extensive damage and haemorrhages.

No major effects are observed on the respiratory system except irritation of mucous membranes. A garlic-like odour has been reported with excess ingestion or inhalation of selenium.

Man is relatively less sensitive to the toxic effects of selenium. Industrial exposure to fumes/dust causes irritation of the mucous membranes of eyes, and the upper respiratory tract. This may also be associated with a metallic taste in the mouth, lassitude, fatigue, dizziness, irritability and pyrexia. In some seleniferous areas, such as Enshi region of China, ingestion of food and drinking water with high selenium content result in chronic human selenosis with symptoms of garlic-like odour in breath, bad teeth, icteroid skins dermatitis, hair loss, diseased nails, gastro-intestinal disturbances, inflammation of mucous membranes, numbness, inappetance, headache, heart burn, convulsions, paralysis, etc.

Selenium sulphide (2.5% suspension or shampoo) is beneficial in cases of seborrhoeic dermatitis, psorariform seborrhoea, seborrhoea oleosa, acne vulgaris, atopic aczema and dandruff. Because of its systemic toxicity, thought washing of hands and finger nails is recommended to remove traces of drug following each application.

Selenium is a cancer- protecting agent. It retards the growth of the certain chemically induced tumours. Epidemiological studies reveal a lower incidence of cancer in areas providing adequate dietary intake of selenium. In cancer clusters and in areas with high incidence of leukaemias, on the other hand, the levels of selenium in drinking water were found to be extremely low.

Silver

The absorption of silver is poor is both the gastro intestinal tract and the lungs, though it can also be absorbed through the skin. A part of the absorbed metal is fixed in tissues.

The silver ion (Ag) has a strong affinity for sulphydryl, amino, and phosphate, groups. It forms complexes with amino acids, purines, pyrimidines, nucleosides, nucleotides, phosphates, proteins, DNA and RNA, and exhibits antagonism to copper. Its excretion occurs primarily through faeces while only traces are found in urine.

Oral toxicity of silver is low in both animals and man. Rare accidental ingestion of medicinal silver compounds or occupational exposure may cause generalized argyria with a slate -grey appearance of skin, hair and tissue due to deposition of silver. No other symptoms are usually observed.

Some silver compounds (e.g. nitrate, allanotoinate, sulfadiazine, etc.) are used for local antiseptic and caustic effects. Slate-blue pigmentation of the elastic fibres of skin is a characteristic feature of chronic silver poisoning. It starts on the face, hands and finger-nails which may later spread all over the body. The pigmentation is usually permanent.

Colloidal silver is carcinogenic. Subcutaneous implantation of silver foil/particles has been shown to induce sarcomas in rodents. The metals does not seem to pose a risk to human beings through contact or ordinary use.

Thallium

The accidental or homicidal ingestion of thallium containing pesticides, rodenticides and depilatory agents is the major cause of its exposure to humans. It is manifested by gastroenteritis, mixed neuropathy (motor and sensory), pain, paraesthesia, behavioural disorders, choreform movements, alopecia, vision defects, and damage to the central nervous system which may ultimately prove to be fatal. The effects have been attributed to the formation of an insoluble thallium complex with riboflavin and interference with the sulfur metabolism.

The absorption of thallium from mucous membranes after

ingestion or inhalation is rapid and substantial. It may also be absorbed on contact with skin. Though a rapid distribution occurs from blood to tissues, thallium is retained longer in blood plasma (as compared to other metals, such as barium, nickel or zinc), suggesting its cumulative nature. Its storage occurs mainly in intracellular fluid where it combines with amino acids and builds up in bones as phosphate.

Competitive replacement of Tl^{+} by K^{+} has been reported. Binding of thallium by ion-exchange in the gastro-intestinal tract constitutes an effective mechanism for its detoxification. The bulk of excretion occurs via urine and faeces within four weeks of ingestion. It may also be eliminated via saliva and skin. An analysis of thallium in scalp hair may provide forensic diagnosis and therapy control.

Thallium has been reported to cause DNA breaks, mutagenesis and carcinogenesis in experimental animals.

Tin

Tin is the widely used mainly for making tinplate and various alloys and compounds. It is an essential micronutrient and the main cause for concern regarding its toxicity has been the development of trialkyl-tin and triaryl-tin compounds having powerful biocidal properties. These are used on growing crops as fungicides and insecticides. They are also used as antimicrobial agents and in marine anti-fouling paints. They can damage crops and, in animals, accumulate in the central nervous system with harmful effects.

Inorganic tin compounds are relatively non-toxic. Trialkyl-substituted compounds are the most toxic organotin compounds. Their exposure results in a severe damage to the central nervous system, and it may cause death from paralysis of respiratory muscles. Its widespread poisoning, from the medicinal use of a tin preparation for the treatment of a staphylococcal skin infection has once been recorded in France. The major cause of toxicity was attributed to the presence of substantial amounts of triethyl tin as impurity in the medicine. Cases of food-poisoning due to tin contamination in canned food and fruit juices have been reported with non-specific symptoms, such as nausea, vomiting, diarrhoea, ataxia, etc. Experimental data from laboratory animals indicates the interaction of tin with calcium, copper, iron, selenium, zinc and other nutrients, causing anaemia, disturbances in cellular functions and bone mineralization.

A brief contact with alkyl tin compounds elicits on irritant effect on respiratory mucosa. Prolonged inhalation of tin dusts causes begins pneumoconiasis. Toxicity may occur not so much from tin as from contaminants (As, Sb, Pb, etc.) in tin compounds.

Vanadium

Vanadium is widely distributed. It is used as an alloying element for steel and iron, in making oxidation catalysts and in colouring agents used in the ceramic industry. Large amounts enter the atmosphere from the burning of some petroleums. It is an essential micronutrient and may be accumulated by some marine organisms to concentrations many times higher than those in the surrounding water. Excessive levels in animals inhibit tissue oxidation and synthesis of cholesterol, phospholipids and other lipids, and amino acids. Such levels may also cause precipitation of serum proteins.

The absorption of vanadium inhaled in particular forms entering the lungs is rapid compared to its oral absorption. Most of the vanadium entering the blood stream is rapidly eliminated with only a small fraction retained and deposited in tissues. Much of the vanadium present in physiological systems is bound to coordinating groups, such as -SH, -SS, -OH, -N, -COO-, and -PO_4. It affects several enzymatic reactions, the sodium pump, oxidation-reduction enzymatic reactions, the sodium pump, oxidation-reduction processes, metabolism of iron, carbohydrates and lipids, bone and teeth formation and hormonal functions. The bulk of ingested vanadium remains unabsorbed and is excreted in faeces, while the remainder is eliminated via urine and bile.

Because of the low toxicity of ingested vanadium, dietary poisoning does not normally occur. Inhalatory exposure to vanadium fumes causes irritation to the respiratory mucous membranes and conjugative, with symptoms of cough, rhinitis, conjunctivitis, chest pain, asthma, bronchospasm, nausea and reduction in the vital capacity of lungs.

Vanadium has been hypothesized as one of the etiological

factors of manic-depressive psychosis (MDP). High dietary-intake interferes with the sodium pump in susceptible individuals, inducing symptoms of MDP. These symptoms are often relieved by the administration of vanadium inhibitors, vitamin C and ethylenediamine tetra acetate (EDTA).

Mice given a 5 mg/ml of vanadyl sulphate for a life-time develop cancer. There is no report to suggest any association between this metal and cancer in human beings.

Zinc

Zinc makes up only 0.004% of the earth's crust. Its most important use is as a protective coating on other metals, particularly in galvanizing iron and steel. It is an essential micronutrient and an essential constituent of alcohol dehydrogenase, carbonic anhydrase, alkaline phosphatase, carboxyl peptidase B, and other enzymes, (Noggle and Fritz, 1986). In most natural waters zinc is found in traces (less than 1 mg per litre i.e., well within the safe limits). Concentrations above 5 mg per litre causes disagreeable taste, In drinking water the level of zinc usually ranges from 0.005 to 1 ppm or mg per litre, but in certain regicns it may exceed upto 7.0 mg per litre. Zinc is generally regarded as one of the less hazardous elements, though its toxicity may be enhanced by the presence of arsenic, lead, cadmium and antimony, as impurities. Toxic effect have been observed from the inhalation of fumes from galvanizing baths. The 'zinc fever' produced is characterized by chills, fever and nausea. Removal from the fumes leads to complete recovery. Zinc chloride fumes have sometimes cause fatal oedema of the lungs. Zinc or galvanized containers are not recommended for food storage but are acceptable for storing drinking water. This is because acidic foods can dissolve enough zinc from the container to cause poisoning. A factor which serve to minimize the risk of zinc poisoning is that it appears to be lost along food chains, unlike methyl mercury or cadmium, for example, which accumulate.

The absorption of ingested zinc takes place in small intestines, particularly through the distal duodenum and proximal jejunum. It is equally well absorbed as oxide, carbonate and free metal, but poorly sulphate or as mixed oxides of zinc, iron and manganese. Metallotheionein plays an important role in the homeostatic regulation of zinc absorption and metabolism.

Absorbed or parenterally administered zinc binds to plasma proteins (albumin, α-2 macro-globulin and transferrin) and is distributed throughout the body tissues. Its highest concentrations are reported in prostate, liver, muscle and kidneys. The metal crosses the placental barrier to reach the foetus. Within the body, the metal is a constituent of a large number of enzymes and proteins and plays an important role in enzymatic reactions, nucleic acid metabolism, protein synthesis, maintenance of membrane structure and function, protection against free-radical damage, replication, transcription and translation of genetic material, and hormonal activity.

Ingestion of large doses causes irritation and corrosive damage to the gastro-intestinal tract, with symptoms of nausea, vomiting, cramps, colic, diarrhoea, fever and fatal shock. Its toxicity may occur from an intake of acidic fruit juices stored in galvanize, i.e., zinc plated, containers where zinc concentrations may reach upto 2g/ litre.

Metal fume fever results from inhalation of more than 5 mg ZnO pe M^3 with a particle size of 0.2 to 1 μm. It is attributed to the irritation of an immune reaction and manifested by sore throat, cough, chills pneumonia, pulmonary oedema, fever, nausea and vomiting.

Deficiency of Zn is known to cause acrodermatitis enterpathica in man and rough skin in animals. The element possesses antiseptic, astringent, styptic and healing properties and finds place in many skin ointments, lotions and dusting powders. Increased urinary excretion and decreased plasma levels are general feature of infectious disease including leprosy.

The carcinogenic properties of zinc have not been fully investigated. Like other chromates, zinc chromate is a carcinogen. Application of hydrocarbon carcinogens causes a decrease in skin-zinc content. The metal is needed for cellular proliferation of existing tumours. Tumour growth is, therefore, retarded in zinc deficiency. The metal does not pose a risk for carcinogenesis in humans. However, a significantly low concentrations of zinc in blood, plasma or serum have been reported in bronchial carcinoma, Hodgkin's disease, lymphoma, multiple myeloma and chronic lymphocytic leukemia. In cases of carcinoma of prostate glands, the organ contained low zinc concentrations but the values wee high in hyperplastic prostate glands. Significantly high zinc levels were found in the liver of carcinoma patients. No comparable increase in the zinc content of kidney, spleen, heart, or pancreas was observed in autopsy samples of these patients.

REFERENCES

Anand, S.J.S. 1978. Determination of Mercury, Arsenic and Cadmium in fish by neutron activation. J. *Radio, Anal. Chem* 44, 101-107.

Goldberg, E.D. 1976 . The *Wealth of the Oceans.* The UNESCO Press Paris, 172.

Lopez, J.N. and G.F. Lee 1977. Environment Chemistry of Copper in Torch Lake, Michigan, *Water, Air and Soil Pollution*, 8, 373-385.

Noggle, G.R. and G.J. Fritz 1986, *Introductory Plant Physiology*, Prentice-Hall of India Pvt. Ltd., New Delhi.

Pillai, K.C. 1985, Heavy metals in aquatic environment. In C.K. Varshney (Ed.) 1985. *Water Pollution and Management*, Wiley Eastern Limited, New Delhi.

Rai, L.C. and M. Raizada, 1987. Toxicity of nickel and silver ions to *Nostoc muscorum* : Interaction with ascorbic acid, glutathione and sulphur containing amino acids. *Ectoxiology and Environmental Safely*, 14, 12-20.

Rai, L.C. and M. Raizada, 1988. Impact of Chromium and Lead on *Nostoc muscorum* : Regulation of toxicity of ascorbinc acid, gluathione and sulphur containing amino acids. *Ecotoxiology and Environmental Safety*, 15, 21-32.

Srivastava, Alka and V.S. Jaiswal, 1989. Effect of Cadmium on turion formation and germinatioin of *Spirodela polyrrhiza L.J. Plant Physiology*, 134, 385-387.

Alloway, B.J. (ed) (1990) *Heavy Metals in Soils*, Blackie and Son.

Bowen, H.J.M. (1979) *The Environmental Chemistry of the Elements*, Academic Press, London.

Fergusson, J.E. (1990) *The heavy Elements : Chemistry, Environment Impact and Health Effects*, Pergamon Press, Oxford.

Waldron, H.A. (ed) (1980) *Metals in the Environment*, Academic Press London.

Bowen, H.J.M. (1979) *The Environmental Chemistry of the Elements*, Academic Press, London.

Fergusson, J.E. (1990) The Heavy Elements : Chemistry, Environmental Impact and Health Effects, Pergamon Press. Oxford.

Harte, J., Holden, C., Schnieder, R. and Shirtley, C.(1991) *Toxics A to Z.* University of California Press, Berkeley and Los Angeles.

Kabata-Pendia, A. and Pendias H. (1984) Trace *Elements in Soils and Plants*, CRC and Press, Boca Raton, FI.

Kazantis, G. (1980) Chapter 8 in Waldron, H.A. (ed) *Metals in the Environment*, Academic Press, London.

Kloke, A., Saurebeck, D.r. and Vetter, H. (1984) in Niragu, J.O. (Ed) *Changing Metal Cycle and Human Health*, Springer-Verlag, Berlin.

Krauskopf, K.B. (1967) *Introduction to Geochemistry*, McGraw-Hill, New York.

Langard, S. (1980) Chapter 4 in Waldron H.A. (ed) *Metals in the Environment*, Academic Press, London.

Manahan, S.E. (1991) *Environmental Chemistry* (5th edition), Lewis Publishers, Chelsea, Mich.

O'Neill, P. (1990) Chapter 5 in Alloway, B.J. (ed) *Heavy Metals in Soils*, Blackie and Son, Glasgow.

Waldron, H.a. (1980) Chapter 6 in Waldron. H.A. (ed) *Metals in the Environment*, Academic Press, London.

7

Toxicity of Pesticides

INTRODUCTION

Like the air we breathe and the water we drink, pesticides are taken for granted. Find a single weevil in a bag of flour, and the seller may land in trouble; one bright-green caterpillar tumbling from a cabbage on the chopping board produces shrieks of horror. Yet the familiar unblemished fruits and vegetables, unpopulated bags of flour, and even mould-free bread can exist only because pesticides are used at every stage of food production. Most of us are unaware, or a least unmindful, of the daily sprinkling, and spraying of pesticides on farms and elsewhere in the environment. Such relaxed attitudes are being shaken by mounting publicity about the widespread use of pesticides and by facts which challenge the comforting thought that 'if the Government approves them, they must be safe'.

We probably could not live without pesticides, but we *certainly* cannot avoid them. Their use is not limited to saving our food crops from destruction or keeping life-threatening pests like malaria mosquitoes and plague-carrying rats at bay. Often they are used for purely cosmetic purposes or as a minor labour-saving device. The result is that our communities, our daily lives, are surrounded by pesticides.

Growing crops and farm animals are sprayed and dipped stored food is protected with pesticides. Most homes have insect sprays, and gardeners keep an armoury of powders and liquids to keep slugs, insects and fungi off treasured crops. Even the timber, paint, and carpets are impregnated with chemicals to stop moulds or insect damage. Fresh flowers from the florist and potplants from the nursery have been treated. The mains water has chemicals deliberately added to kill pathogens, and a cocktail of both accidentally and deliberately added pesticides washed from farms and factories on the way. Even the family doctor prescribes specialized pesticides in the form of antibiotics and other drugs to kill the bacteria, lice, worms, and fungi which undermine our health.

An important reason for thinking more carefully about the pesticides in our lives comes from reflecting on what they do. All niceties aside, pesticides are useful because they kill. The word pest comes from the Latin for 'Plague' and the *cide* ending from the Latin *Caedere*: 'to kill'. Whether it is harmless ants in the kitchen, whitefly on house plants, crop-destroying caterpillars, yield-reducing weeds and fungi, or bacteria that threaten human health, pesticides are designed to disrupt the chemistry of life and to kill.

There are many similarities in the biochemistry of plants, insects and human. There should be no surprise in discovering that a pesticide designed to kill plants like dandelions may also harms or kill us. The common garden weedkiller paraquat is extremely poisonous if swallowed and there is no known antidote. Chemicals like the insecticide malathion, or warfarin, the rat and mouse bait, target organisms much closer to humans than the dandelion. Malathion works by blocking on enzyme which controls the sending of nerve signals around an insect's body. The very same *enzyme* has the same *job* in the human nervous system. We should expect such chemicals to be harmful to us also, and perhaps in amounts similar to those used against pests.

Many of these chemicals are highly complex and unique structures whose method of action against pests is sometimes only partially understood. Many synthetic chemicals are totally new substances, invented chemicals that have never been known before. It is not surprising that their effect on other organisms in the environment are little understood, and therefore mostly unpredictable, when they are first released onto the market. Most of the new pesticides being sold are not single chemicals but mixtures of two, three, or more active ingredients, usually mixed with chemical solvents, or surfactants, to help in squirting them through a nozzle and to make them stay where they settle. Of more concern, as the number of these new pesticides increases, is the lack of knowledge about how they will react with each other in the environment, and about the unplanned effects that will result.

While some pesticides break down harmlessly within hours or days of application, others are designed to stay active for moths while they wait for their target organism to appear. Some pesticides work by lying on the surface but other designed to get inside the plant or animal they are protecting. Many do not break down rapidly, or are applied at excessive doses and persist as residues in and on foods and other products. This contamination of things we eat or handle supplies us with a steady dose of substances that are odourless, colourless, and tasteless, but deadly poison.

Large numbers and large quantities of pesticides have been around for a remarkably short period — since 1950 when the petrochemicals industry took off on a sea of cheap petroleum resources. Only in the last 15 or 20 years have scientists realised, to their horror, that chemicals they thought safe at first have been ticking away like time bombs throwing up disease many years after their introduction. Most safety testing is concerned with short-terms or acute illness, usually with large doses of the test substances in mind. But we have learned the hard way that substances like carcinogens show no effect at all until perhaps 30 years after they have entered the body. These chronic effects are difficult to predict and very hard to prove. Carcinogens need be present in the minutest of quantities, perhaps only a single molecule, to have their effect. Pesticides like the insecticides aldrin and dieldrin, or the anti-fungal wood preservative lindane, which is used to protect timber, can cause mutations or cancers years after exposure.

Many relatively safe pesticides can be harmful if abused if they accumulate in the food chain. There are many examples of wildlife dying or suffering disrupted reproduction because of pesticides used to kill other organisms. Honeybees, which are vital to successful cropping, are often accidental victims of sprays aimed at harmful insects like grubs and beetles on food crops. Rachel Carson's chilling book *Silent Spring* chronicled the ecological disaster of DDT, at the time considered the most successful and widely used pest-controlling substance. It was cheap, apparently safe to humans, and an effective killer of insects. However, DDT was shown to resist being broken down once it had done its job. Organochlorine insecticides like DDT accumulate in the food chain because they build up in the fatty tissues of organisms that are then eaten in quantity by higher organisms, thus moving up the food chain and threatening fish birds and higher animals like humans. Through such an ecological multiplier effect, dieldrin was blamed for the losses of peregrine falcons in the UK during the 1960s, and, far away in Antarctica, penguins have been found to have enough accumulated pesticide residue to cause their eggshells to be soft and break.

Agriculture uses 83% of the total pesticides applied in the UK, but they are also a normal part of urban and industrial life. Most pesticides are applied but out of sight so we rarely consider them: on playing fields, in factories, along railway embankments, and in parks. But it is remarkable how we also use them in our own homes apparently oblivious to the risks involved and, often, without knowing that we are handling pesticides. Little wonder when household chemicals are advertised with photographs of people spraying their gardens without even rubber gloves to protect themselves from skin contact. This is far cry from advertisements for farm chemicals where farmers are shown dressed in as much protective clothing as astronauts. This difference reflects new laws protecting workers, but there are no similar controls to protect the home user.

Choosing to spray the garden roses is one thing, but how many people choose to have pesticides in their new carpets or in the wood for their bookshelves? Pesticides are routinely included in many household paints and wallpaper pastes, but this is not mentioned in advertisements, which entice with fancy patterns and fashionable colours, or on the labels attached to these goods. There is no legal requirement to inform shoppers when pesticides have been used in household items.

But even knowing what chemicals are used, and where, does not give enough information for wise decisions. Not only is there little publicity of the specific dangers in using pesticides, but the results of tests showing their effectiveness and their toxicity to ourselves and wildlife are also difficult to discover because the data are confidential between the manufacturer and the government.

The testing procedure that new chemicals have to pass before being approved are guarantee of safety to man and the environment. Although present procedure for approval are more rigorous than ever, there remain doubts, about their ability to screen out unacceptably harmful chemicals. The more of these chemicals that are used, and the longer they are in the environment, the greater grows our experience of harmful mistakes. Many chemicals are approved, or allowed, continued use, even their safety has been questioned, on the basis of incomplete or unreliable data. Such uncertainly is rarely admitted by the authorities, who encourage consumers to believe that the each approved chemicals has equal amounts of indisputable scientific data behind it.

The most glaring problem for consumers trying to reach opinions or make decisions about pesticides is our lack of information. We live in a soup of chemicals but have insufficient information about where they are, what they are for, the risks involved, or how to use them with greatest safety. Products treated with pesticides are neither labelled nor dyed pink, so we cannot tell the difference between the risks we pay for and the safety we pray for.

In a market economy, supply and demand are allowed to dictate what is sold on the basis of three rather dubious assumptions. First, that consumers are fully informed; second, that they are able to make free and unconstrained choices when shopping; and, finally, that they use their wisdom and personal income to make the most efficient purchases possible. In general, this is a rather shaky basis for complacency about the economic or social wisdom behind what is sold and bought. Where pesticides are concerned, these assumption are less valid than elsewhere in the market place.

There are no other substances found in or on the food or other goods we buy, or applied to the field, roads, and buildings or our environment that are treated with similar secrecy and lack of public consent. There is no other product marketed which is consumed involuntarily by the entire population. With the exception of nuclear energy, there is not other product with the same potential to harm both human health and the future environment of the earth.

WHAT IS A PESTICIDE ?

The term 'pesticides is usually restricted to chemicals. Sulpher was used as an insecticide in 1000 BC and both the Greeks and

Chinese used arsenic. Even household fumigation has an ancient precedent: The Greeks burned thyme to rid their homes of insects.

Pesticides conjure up images up complex chemicals with daunting names. Take the herbicide alachlor. It is known to chemists as 2-chloro-2, 6-diethyl-N-methoxymethylacetanilide, and has the formula $C_{14}H_{20}CINO_2$. There are simple chemicals like sodium chloride (NaCl), otherwise known as common salt, which does simple jobs like killing moss on paths, or copper sulphate (CuSo), which is used in fungicides sprayed on leaves. Life used to be simpler.

The early pesticides were either natural substances or simple inorganic compounds. Dung, urine, ashes, mud, and various plant extracts have all been used. (Stalaman and Harder, 1957.) By the middle of the nineteenth century natural extracts were being used as insecticides. Extracts from the derris and pyrethrum plants were the first example, and both are still used today. Derris powders contain the substance rotenone from the roots of the plant *Derris elliptica*. The properties of derris were known to the Chinese long before Europeans started extracting it. The pyrethrins are a groups of six chemicals found in the flowers of *Pyrethrum* (*Chrysanthe mum*) *cinerariaefolium*. The flowers were originally grown in west ern Asia but are now cultivated in Africa and South America. Both are among the small range of pesticides considered acceptable by the Soil Association.

The wide availability of natural pesticides should remind us of an important fact — pesticides can be both useful and safe. At issue is the safety of certain pesticides both as toxic substances and in how they are used.

More sophisticated use of inorganic chemicals developed in the mid nineteenth century. Sulphur washes were the first deliberate choice of fungicide and were effective against diseases of fruit trees. Copper arsenite was tried against beetles in potato crops in the US, and the famous mixture of copper sulphate and lime, which came to be known as Bordeaux mixture, saved the European vine industry from destruction by mildews between 1880 and 1890. Most of these early pesticides were discovered by accident. Copper sulphate and lime mixture was originally used to make the grapes growing along roadsides taste so unpalatable that they would not be pilfered. An observer noticed that vines which had been washed with the mixture were free from the dreaded grape downy mildew (*Plasmopara viticola*), which first spotted near Bordeaux in 1878, was spreading rapidly and destroying the European vineyards. Before this dramatic chance discovery there had been no protection against the disease and the entire wine industry was under threat. Bordeaux mixture afforded another chance discovery when it was noticed that splashes *beneath* the vine killed some plants but did not harm others. This led to the development of herbicides.

It was a time of realisation and experimentation as the potential of pesticides unfolded. Scientific discoveries in the bacterial and other biological causes of human and plant diseases helped focus experimental on the killing of specific pest organisms. This was the era of Pasteur and Lister. Great efforts went into developing pesticides that would kill pests without damaging valuable crops. Enthusiasm went untempered by a sound knowledge of either chemistry or toxicology. Since common poisons like arsenic and cyanide were to kill pests, they were used widely, even on food crops. Poisonings from zealous applications and the residues left on foods inevitably followed. By 1915 of the first public concern with harm to humans from the careless choice and application of chemicals struck home when the poisoning of farm workers became obvious. The search began for safer alternatives.

This was the beginning of a major shift into organic compounds. Life is full of ambiguous words like 'natural' and 'farm fresh'. 'Organic' is another. It originally signified 'living' substances of animal or plant origin as opposed to earth materials which were inorganic, but developments in analytical chemistry then allowed scientists to copy the constituents of living things in the laboratory.

Now, organic chemistry is simply the study of compounds which contain carbon atoms, irrespective of whether they are found in living organisms or have been artificially synthesized. The consumer should note the distinction between the 'organic' in organic farming and that in organic pesticides.

The skills necessary to make synthetic organic compounds improved rapidly. New chemicals came in families. As the pesticide properties of new synthetic molecules were discovered, chemists set about 'tweaking' the edges of these molecules to make small changes with possible new and useful effects. Chemicals would modify known compounds by adding different atoms alike those of metals or halogens (chlorine, iodine, bromine, fluorine) to the basic organic structures of rings or chains of carbon atoms. This could be done in hundreds of ways and meant testing these totally new compounds in the hope of finding some pesticide activity. Most would have been useless, but the tiny proportion of hopefuls would be developed further as potential commercial products.

During the 1930s, fungicides like the dithiocarbamates and chloranil were developed, and they, or derivative chemicals, are still used today. By 1944, the pesticide that became both the most significant breakthrough for synthetic chemicals and the most notorious environmental poison had been patented. Chemists gave it what they term a 'trivial' name — dichlorodiphenyl-tricholoroethane — but it is more widely known as DDT.

DDT was the first synthetic organochlorine insecticide but was soon followed by some equally notorious relatives: aldrin, dieldrin and endrin, collectively known as the 'drugs', Organo-phosphorus insecticides like parathion and malathion were developed out of the nerve-gas technology which provided the most horrific weapon of the First War, and have the same reputation for being highly poisonous. The carbamate insecticides first appeared in 1947 and became the successor of DDT. The still widely used phenoxyacetic herbicides were developed in Britain in 1943, followed in 1958 by the bipyridylium herbicides, including diquat and paraquat. Later, families like the benzimidazoles, pyrimidines and, of course, antibiotics joined the still-growing list of new chemicals conceived by the giant international industry called agrochemicals.

How Pesticides Work

Living organisms are complex. Apart from the vast number and variety of chemical reactions going on all the time just for a single cell to be 'alive', the cells that make up living organisms contain a large number of structures, or organelles, that are necessary for them to tick over, grow, reproduce or move. The nucleus contains the chromosomes — the genetic code telling the cell what kind of cell it is, its shape, function, colour — in fact, everything to do with being that particular cell. There are also hollow, sausage-shaped mitochondria that store and control the release of the cell's energy stores. The Golgi apparatus is a small spherical object with the job of storing and sorting the proteins made elsewhere in the cell before sending them out to wherever they are needed in other parts of the body. Plant cells have unique organelles, the chloroplasts. These contain the green pigment chlorophyll and are arranged in layers to catch sunlight. The chloroplasts are where light energy is converted by photosynthesis to sugar which serves as an energy source for the plant's activities.

Most pesticides work by interfering with one or more of the life processes which occur within these organelles inside cells. The most common processes affected are described in more detail in the following text.

Blocking Photosynthesis

In photosynthesis, the creation of basic biological energy occurs. All life on earth depends on this process. Put simply, carbon dioxide and water are converted into sugars which are available as an energy source. In the process, oxygen is released, which is one reason why preserving the world's vast rainforests is crucial. The chemical formula for this reaction is:

$$6CO_2 + 6H_2O + \text{Light energy} \rightarrow CO_6H_{12}O_6 + 6O_2.$$

There are two steps in photosynthesis. The first is capturing light energy and converting it into chemical energy. The energy from sunlight is absorbed in the green chloroplasts of leaves and used to split water, releasing oxygen and energy. This energy becomes attached to a special energy-transfer chemical found throughout living things and called NADP. Its chemical name is nicotinamide adenine dinucleotide phosphate.

The second step is the transfer of the energy attached temporarily to NADP to a more versatile storage form. This is a high-energy phosphate bond in a substance called adenosine triphosphate, or ATP. These bonds take a large amount of energy to create and, once formed, can be stored or shipped around the cell or other parts of the organism to where energy is needed to power biological reactions. They work like rechargeable batteries, which can be used to power a wide collection of useful gadgets from torches and food-blenders to computers. ATP is one of the most important substances in living organisms. Photosynthesis creates portable ATP energy from the energy of light.

This light energy, now in the combined form of $NADPH_2$ and ATP, is used to convert the carbon in carbon dioxide into carbohydrate carbon as glucose ($C_6H_{12}O_6$). This sugar is water soluble and can be moved around inside the plant from the leaves to any other part where it is needed, or it can be converted to insoluble starch for storage (roots, tubers, fruits). Any pesticide that disrupts the energy-transfer system thus blocks the process of photosynthesis and halts the plant's vital, and only, source of energy for growth.

Interfering with Engery Release

For a cell or organisms to move, grow, form reproductive cells, or heal injuries, it needs a energy to perform the metabolic processes involved. Plants or algae that photosynthesise use the sugars, or their storage from as starch, for energy. Some plants also create stores of fat that they can mobilise for energy. (Our kitchen vegetable oils come from the energy stores of fruits or seeds—soya, sunflowers, and olive oils.) Insects, fungi that thrive on living or dead plant material, and higher animals use sugars, fats, or proteins for their energy. Even with carnivorous animals, or omnivores like humans, all biological sources of energy can be traced back to photosynthesis. We can use the proteins and fat in beefsteak for energy, but the free-range cow gets energy from the sugars and starch in the grasses and other plants that make her diet. The basic processes for releasing energy are same in all living things.

Sugars like glucose release their energy in two steps. The first, called glycolysis, splits the sugar into two smaller molecules. This requires some energy which is provided by the same ATP energy store created during photosynthesis. Several times as much ATP is produced in the next stage of releasing the energy from the sugar. Glycolysis can occur without oxygen, which is what happens when sudden bursts of muscle energy are needed in running a marathon race. The sugar is incompletely 'burnt', with lactic acids as the end product. Another example is the fermentation of grape sugars in wine or added sugar in beer and lager, producing alcohol as the end product. The fact that alcohol still contains energy from the original sugar is demonstrated by the fact that it will burn, giving off heat.

In the presence of oxygen, the second step breaks the smaller molecules light to carbon dioxide (CO_2) and water, with the release of more energy. This happens in a complicated cycle of reactions known as the citric -acid or Krebs, cycle. In this cycle a single molecule of sugar like glucose is broken down, and 38 molecules of ATP are produced. This mass of ATP is used in the thousands of chemical reactions which keep organisms growing, moving, reproducing, and repairing damaged tissue.

Fats and oils are broken down to their component fatty acids that can be 'burnt' in a similar way to release energy, once broken down to smaller molecules, the 'burning' of fatty acids follows the same citric-acid cycle as sugars. Proteins can also be used for energy as a last resort when fats or sugars are not available. This clearly is a last resort because, unlike sugars and fats, the principal role of which is energy storage, protein is a building material of structural tissue like muscle, tendon, and skin. When protein is 'burnt', it is first broken into the individual units, or amino acids, which make up the large protein molecules, and these are fed eventually into the same cycles that sugar and fat go through. Both fat and protein 'burning' produce ATP.

These processes in living organisms are more complex and varied than are pictured here, but the general picture is true. In each energy-release mechanism, whatever the source of energy used, are numerous interconnecting steps, many of which are controlled by specific chemicals called enzymes. Any pesticide which disrupts these cycles can reduce or stop the release of energy from stores. Without energy, all normal functions of cells and organism stop, and death follows.

Blocking Nerve Signals

Hormones travelling in blood, or nerve signals channelled through nerve fibres, control every action of our bodies, whether conscious or unconscious. The nervous system allows the detection on stimuli like light and pinpricks, and control reactions like jumping up after sitting on a drawing pin. Conscious activity, like turning the pages of this book, or unconscious activity, like the constant beating of the heart, are also controlled by nerve signals.

The nervous systems of insects and mammals have much in common and all work on the same principle. They consist of nerve cells, or neurones, which have a body and a long tail called the axon. The axons are like the wiring inside a house or pocket radio; they carry the electricity from the mains box to every room and appliance.

Nerve signals are small electric currents that travel along the axons from neurone to neurone and can move at about 270 mph. You may not be surprised to learn that ATP is again necessary to power the creation of the electric current.

When two neurones meet and the signal has to pass from one of the other, a connection has to be made, just as in electrical circuits. This connection is called a synapse. The same need for a connector arises when an axon reaches a muscle it controls. The connector is a chemical. There is a small space between two connecting axons or between a terminal axon and a muscle cell. When the electrical signal reaches the end of an axon, it stimulates release of a chemical transmitter into the space. This is most often a chemical called acetylcholine. The acetylcholine 'stimulates' an electrical charge on the second axon (or the muscle fibre), and the electrical wave continues its travel along the next neurone or into the muscle fibre. Once the chemical transmitter has passed its signal across, it is inactivated by a special enzyme called cholinesterase. If it remained in the space, it would block any response to any further signals.

Pesticides can work at two points to disrupt the sending of nerve signals. First, they can stop the creation and movement of the nerve signal. Second, they can destroy the cholinesterase enzyme, which prevents nerve signals going to or from the brain, and stops them connecting with the muscles they control. The result can be paralysis or erratic and incoherent nerve signals. Death usually follows. Several groups of pesticides are known to work through anticholinesterase activity, e.g. carbamates and organophosphates.

Damaging the Central Computer : DNA and RNA

In the center of almost all living cells (mature red blood cells are one exception) is a nucleus that contains the chromosomes carrying the genetic materials, or genes, of the organism. Chromosomes are really the central computer of living things. The chromosomes are made of DNA, or deoxyribonucleic acid, and genes are segments on chromosomes. In human there are 46 chromosomes, 22 matched pairs and two sex chromosomes (X and Y). Women have a pair of Xs and men have and X and a Y.

Chromosomes are duplicated in cell division. In growing

tissue, cells divide repeatedly, making exact copies of each other and increasing the number of cells and the total size of the growing organism. This division is called mitosis. When sex cells, sperm and ova, or eggs, are made, a different division called meiosis occurs. This results in only half the parent cell's number of chromosomes in each sperm or egg cell. At fertilization, the chromosomes in the sperm and egg combine, making up the original number of pairs in the parent.

In cell division and in the copying of the chromosome material, there is plenty of opportunity for 'mistakes' to occur. These are called mutations. Most are disasters for the organism, and cells with mutated chromosomes usually die. Only a minute number of changes leave an organism better adapted to the world. Some mutations occur as errors in the complex copying procedures, but others are due to the damaging effects of radiation or of chemicals in the environment. There are some pesticides that aim to interfere with the duplication of DNA, and some that accidentally result in serious changes to the chromosomes. Both are known as mutagens (causing mutations) or teratogens (causing malformations to the developing foetus). Carcinogens are similar chemicals in that they also disrupt genetic material and cause cells to grow rampantly, without the normal guiding and controlling mechanisms written into healthy genes.

The second important part of the computer is RNA, or ribonucleic acid. This is similar to DNA in structure. RNA is concerned with the manufacture of proteins in cells. Proteins exist in a very large number of shapes and sizes, with functions more varied than simply providing the bulk of muscle. The enzymes and hormones which regulate most body processes and reactions are proteins.

There are several pesticides, especially among the fungicides, which work by disrupting protein synthesis. The synthesis of particular proteins or the blocking of RNA, which stops all protein synthesis, may occur. This usually disrupts vital enzymes, blocking normal functioning, and death follows.

Interfering with Growth Regulators

Plants, Insects, and higher vertebrates like humans all have their growth regulated by hormones. In plants, hormones like indoleacetic acid and the gibberellins regulate cell growth. They ensure that leaves and stems grow towards the life-essential sunlight, as well as ensuring that the rapid growth of new cells in growing shoots and roots maintains strict order. There are also hormones which regulate the time of flowering and leaf-drop in the autumn. Insect hormones control growth in larval stages, the moulting, and the formation of the waxy, protective cuticle, or shell, of adult insects, as well as the development of sexual organs. Human harmones of comparable importance are thyroxine, which controls the pattern and rate of growth, and the sex hormones, testosterone, progesterone, and oestrogen. Such hormones control, among other things, the masculine body appearance and the preparation of the womb for nurturing the growing foetus.

Chemicals which mimic or inhibit such hormones can block growth or reproduction in pests. Some herbicides work on the mimic principle, and there is increasing interest in developing more hormone-mimic insecticides rather than poisons which can have serious environmental effects. Hormone mechanisms ofter possibilities for targeting of pesticides very specifically, as well as for greater environmental safety.

Classifying Pesticides

There are many forms of classification. For example :

- **Natural or synthetic** Natural pesticides such as derris and pyrethrum can be effective while remaining free of many of the more serious environmental hazards of some synthetic chemicals like DDT.
- **Organic or inorganic**
- **Surface acting or systemic** Surface-acting chemicals remain on the surface of plants, soil, or other substances and objects, and go to work when the organism to be killed comes into contact with them. The systemic are absorbed into the tissue of the plant or animal being protected and spread throughout the body, particularly in plants. (Antibiotics are one example of a systemic pesticide.) The advantages of systematics are obvious: they cannot be washed off cone absorbed through the outer layer of cells; they can reach areas such as they rots

which surface leaf sprays cannot; they can kill organisms (e.g. fungi and bacteria) which have penetrated inside the plant; they can eradicate pests that start to grow or eat their way in, e.g. germinating fungal spores, fungi, caterpillars or sap-sucking insects. But systemic pesticides also have the disadvantage that they are impossible to wash off treated crops.

- The most useful means of classification are by **function** and by **chemical family**.

Function

The majority of pesticides used today are insecticides, fungicides, or herbicides, but there are many other uses.

PESTICIDE	USED TO CONTROL
herbicide	fungi
insecticide	insects
nematicide	nematodes
rodenticide	rodents (mice, rats, etc.)
bactericide	bacteria
arachnicide	spiders
algicide	algae
acaricide	mites and ticks
molluscicide	molluscs
avicide	birds
slimicide	slimes
piscicide	fish
disinfectant	general bacteria, fungi
growth regulator	growth of plants
attractant	insects by attraction
repellent	flies, fleas, moths, etc.
defoliant	leaf drop
chemosterilant	insects by sterilization
desiccant	plants by drying leaves

Knowing the function of a pesticides gives many clues about its nature. For example, redenticides kill rats and squirrels, which are mammals, as are pet cats and dogs, and indeed, as are we. Rodenticides are likely to be dangerous to *our* health.

Insecticides often work by blocking the nervous system or energy-transfer mechanisms. It is likely that exposure to insecticides will affect out nervous system or metabolism. All insecticides should be used very carefully around humans.

Attractants, on the other hand, have no insecticidal function but serve only to bring the target pest into contact with an insecticide. So an attractant is not likely, in normal use, to be harmful to humans.

Chemical Family

Chemicals are allocated to families on the basis of their formulae, and members of one family often have properties. The family type gives an idea of the individual chemical's function, behaviour in the environment and toxicity to humans. Media reports and articles on pesticides tend to use chemical family or group classifications as much as the trade or chemical name for specific chemicals. Learn about a family, and you will have a good idea how either new a chemical or an unfamiliar brand of pesticide may behave.

There are many possible families to include in a classification and no universal agreement. Many chemicals can be placed in more than one family. The classification used here gives an idea of the larger families into which the most commonly used pesticides fall, and it covers the main types of chemical, where health and environment effects are of interest. The families don't account for every available pesticide because some don't belong in any of these families, but the omissions are the exception. The main families are described in greater detail below.

CHEMICAL FAMILIES OF PESTICIDES

The following are the larger families into which the most commonly used pesticides fall. Some chemicals do not belong to any of these families, but they are the exception.

ORGANOCHLORINES

These are based on a benzene ring with one or more chlorine atoms attached, either to the ring or to side chains. The best-known member is DDT.

DDT related compounds were developed rapidly as the insecticidal power of DDT became legendry, but few attempts produced insecticides as powerful as the parent, DDT. Example include methoxychlor and DDD, DDT is now banned in the UK, but it still appears in food residue tests at levels suggesting that it is being applied illegally on some crops of lettuce, mushrooms, and tomatoes.

Hexachlorocyclohexane (HCH) or lindane (BHC) : More soluble than DDT, lindane, otherwise, known as BHG, is widely used in insecticide mixtures in the UK. It is used in soluble soil treatments as well as for fumigating or spraying against a wide range of insects. Lindane is under review by MAFF. Common products include Boots Ant Destroyer, Murphy's Gamma BHC Dust, and shampoos and creams against human lice and scabies.

Cyclodienes — the 'drins' Chlordane and heptachlor were early entrants to this group, but aldrin and dieldrin are the best-known members and are named after the two scientists who discovered the creating reaction — Diels and Alder. Heptachlor has a minor use in the UK in worm-controlling chemicals like Mascot Wormort and Parker's Wormex. The cyclodienes are usually complex, three-dimensional combinations of fused pentane, or five-carbon rings.

Aldrin, endrin, and dieldrin have similar structures and properties. Like DDT, they are soluble in fats and do not degrade easily in the environment. They are effective insecticides, working by contact, and hence are used as a soil insecticide or are applied to surface visited by flies and ticks and on fabric where months may lay eggs. Total bans on the use of aldrin and dieldrin in the UK have been in force since 1989, while endrin has been banned for much longer. Because of their persistence in the environment, their residues will continue to be found in the food chain for some years to come. Despite being an established carcinogen and highly persistent, chlordane is still permitted in the UK for use on public-amenity turf against worms.

Other organochlorines Chemicals like pentachlorophenol, which is similar to lindane, are used in herbicides, fungicides, and wood preservatives. Several important fungicides like captain and captafol are based on the benzene ring but contain sulphur as well as several chlorine atoms. The fungicide quintozene is also similar to lindane but has nitrous oxide in place of one of the chlorine atoms on the ring.

Mode of action Despite being among the earliest of mass-produced pesticides, it is remarkable humbling that we still know very little about how organochlorines act. Clear hints are given in the writhing, jerking death throes of insects under their influence. It seems they interfere with the transmission of nerve impulses, either eliminating or jumbling the signals. Without clear nerve signals such vital functions as breathing cannot occur, and general body muscles twitch erratically and uselessly. How the sulphur-containing fungicides in the group work is unknown, but sulphur has long been known to be toxic to fungi.

Health and environmental effects Organochlorines vary in their acute toxicity for humans from an LD_{50} in rats of 3 mg/kg for endrin to 457 mg/kg for DDT. A few drops of endrin or a tablespoon of DDT may kill an adult human. Acute poisoning causes dizziness, nausea, and twitching of muscles in arms and legs, followed by tremors and convulsions. Breathing may speed up at first but can stop completely. Chronic poisoning may arise because of the tendency of the group to accumulate in fatty tissue within the body, causing kidney and liver damage. Long-term build-up in body fat can also lead to sudden release and symptoms of acute poisoning if fat reserves are mobilized, for example, with sudden depletion of excess fat during dieting or a period of psychological stress.

This group has also been found to contain carcinogens, mutagens, and teratogens, which is one of the main reasons they are banned, or at least severally restricted, in many countries. DDT, aldrin, dieldrin, heptachlor are recognised carcinogens and teratogens. Lindane is the only one of the group approved for use in the EG, but it is under suspicion as a carcinogen and mutagen. It is suspected of causing leukaemia among the factory workers who make it.

Organochlorines are broad-insecticides, meaning they kill a large of insects which may include harmless or beneficial insects as well as the target pest. They also harm fish, birds, and mammals.

This entire family is distinguished by its persistence in the environment. The chemicals are not easily broken down in soil or water so they accumulate. DDT has a half-life of 20 years in soil. One headache with long-term persistence is the speed with which insects develop resistance. Organochlorines also have in the awkward habit of accumulating in the fatty tissue of insects, birds, fish, and animals. This causes the food-chain multiplier effect, in which doses too small to kill organisms accumulate in body fat and become concentrated as animals higher up the food chain eat large numbers of these lower organisms as their food. DDT travels up the chain through algae and plankton, insect larvae, small fish, and rodents, through birds and larger mammals to still larger birds and animals, concentrating at every step. Rare birds to prey still die each year in the UK because of this effect. In February 1990, two of 11 reintroduced red kites were found dead from endrin poisoning even though endrin was banned a decade ago.

HALOGENATED HYDROCARBONS

Unlike the organochlorines, this groups is based on straight chains of carbon atoms to which are attached not only chlorines but also other halogen atoms like bromine and fluorine. The group includes herbicides and nematicides as well as insecticides. They are often used as fumigants because they vaporise easily and can thus diffuse through stored gain, glasshouses, and the soil.

Bromide group : Among the early and effective fumigants are methyl bromide and ethylene dibromide. These are toxic to mammals, and methyl bromide, in particular, is extremely toxic to humans. Now banned in Europe as a means of killing insects in stored fruits and vegetable, the toxicity of ethylene dibromide has been used as a justification for food irradiation as an alternative pest control.

Dibromochloropropane (DBCP) is a largely phased out nematicide fumigant used in the 1950s on Third World crops of tobacco and pineapples.

Chlorinated group : Two pesticides against nematodes come from propane and propene gases with the addition of chlorine — dichloropropane and dichloropropene. Propionic acids is the starting point for two useful herbicides, especially active against grasses. Dalapon is a translocated herbicide useful against grasses like couch, and its cousin, trichloroacetic (TCA) acid, also hits grasses but is less effective. It works in the soil rather than by travelling up inside the weeds.

Mode of action The nematicides in this group attack nematodes by being lipid (fat) soluble and penetrating the outer protective coat of these worm-like organisms which live in soil. Once inside, they act like narcotics and depress the activity of the nervous system. The herbicides disrupt and block normal activity in the plant cells they contact by precipitating proteins and therefore deactivating the important cellular enzymes which control metabolic reactions. Pantothenic acid, an important B vitamin in both human and plant nutrition, is one enzyme destroyed. Damage can be widespread.

Health and environmental effects. The volatility of halogenated hydrocarbons and their solubility in fact make their entry into human bodies relatively easy, and they become trapped in fat tissue. Like chloroform, which is nothing other than trichloromethane ($CHCl_3$), these chemicals anaesthetise the central nervous system. One old pesticide (chlorpictin) was made from chloroform. The chlorine and bromine atoms are particularly toxic to the peripheral nervous tissues. Kidney damage can occur in chronic exposure. They are active alkylating substances, which means they can damage DNA, causing mutations and cancers. Ethylene dibromide is classified as a potent carcinogen and mutagen with the ability to penetrate protective clothing, rubber, and with ease, human skin. DBCP has been proved to be a sterilant of factory workers and carcinogen.

While the herbicides are reasonably specific to such grasses as couch and wild oats, the soil fumigants have a broad-spectrum effect, killing of many harmless and even beneficial organisms in the soil. Some, like dalapon, are also mobile within the soil and can be washed into ground water. Wide use of DBCP on cotton in the US forced closure of underground wells.

CARBAMATES

This is an interesting group which developed from the work on organophosphorus insecticides which also act by disrupting nerve signal transmission. They are derived from carbamic acid, which does not exist in the free state.

The Geigy Company introduced the first effective insecticidal member of this group —Isolan—in 1951. It is water soluble and effective against aphids and houseflies, but it is so toxic to mammals that it was withdrawn. New carbamates based on a pyrimidine ring, that is, a benzene ring with two nitrogen atoms replacing two of the carbon atoms, were found to be safer. Pirimicarb was less toxic but a good systemic aphicide. Later, phenolic rings were used, producing carbaryl, also known as sevin. Carbaryl is a broad-spectrum insecticide used on vegetables, fruit, cotton, and turf. If was an early replacement for DDT because it breaks down and leaves no harmful residues.

Other insecticidal carbamates include cabofuran, propoxur, aldicarb, and methomyl. One chlorinated carbamate which was developed as a pre-emergence herbicide is chloropham, and a relation, barban, was used on established weeds.

Thiocarbamates Sulphur-containing carbamates, known as thiocarbamates, like Eptam (EPTC), are pre-emergence herbicides used against grasses and broadleaf weeds like at fat hen and chickweed in potato crops.

Out of the vulcanised rubber industry emerged the first pure organic fungicides — the dithiocarbamates. The first of the bunch was thiram, which is still used. The most important family within the groups are the ethylenebisdithiocarbamates (EBDCs), which are widely used fungicides. The first, nabam, is converted into its siblings zineb and maneb by reaction with zinc and manganese salts, respectively. The EBDCs were the subjects of an urgent review by the government during 1989 after they were dropped from the US lists on grounds of carcinogenicity. They remain on the approved list in the UK. Ring-structured carbamates were developed with varying specificity for grasses or specific broadleaf weeds like dock and bracken. Triallate was a sulphur-containing carbamate aimed at wild oats. Asulam appeared in 1965 as successful treatment for the bracken which is still spreading over the upland hills of the UK.

Mode of action. The carbamate insecticides are anticholinesterase in action (see Appendix A) and disrupt the nervous system. This means disordered nerve signals, and where muscles are concerned, there are convulsive jerkings and eventual death from respiratory or heart failure.

The fungicides may work through several different actions. The EBDCs seem to break down readily into toxic substances that react that important thiols compounds in the fungal cells, or they may bond strongly with trace elements like copper in the cells, starving the cells of the use of the copper. Other members block energy transfer in the Krebs cycle, which is crucial in the metabolism of cells.

Some of the herbicides work by blocking the photosynthetic reaction. The absence of photosynthesis causes death as soon as stocks of carbohydrate run out. Others disrupt the process of the cell division, mitosis.

Health and environmental effects. Aldicarb (Temik) is one of the toxic pesticides available, with an oral LD_{50} of 0.93 mg/kg for rats and an ADI for humans of only 0.0005 mg/kg. It is also toxic to wildlife and fish. The group is water soluble and easily absorbed through the skin or on breathing. The signs of serious poisoning are typical of anticholinesterase pesticides, and chronic effects include depression, nerve damage, and poor memory. They break down into ethylene thiourea (ETU) in sunlight, and in food processing and cooking. ETU is an animal carcinogen, producing thyroid cancers. Although the EBDCs are not absorbed into the sprayed plants, commonly potatoes and wheat in the UK, the breakdown products, or ETUs, are absorbed. Residues are found in lettuce, spinach, and fruit like apples and pears.

With the exception of aldicarb, the group break down readily in the environment. They can affect such non-target organism as fish and earthworms.

HETEROCYCLIC COMPOUNDS

This is a large group of successful herbicides based on a ring structure containing both the usual carbon atoms and nitrogen atoms. Many are so effective that they are used as total herbicides on sites like car parks or around buildings, where they destroy everything green and growing. The same herbicides in smaller doses are more subtle, killing only broad-leaved weeds when used on crops like sugar can and asparagus or around soft-fruit bushes like gooseberries. Grass family crops, including sugar cane and maize, are unaffected by members of the group, like atrazine, because they contain special enzymes which break down the herbicide as soon as they get into the plants.

Triazines. These are soil-acting herbicides that are taken into the weeds through the roots and kill by destroying the photosynthetic function of the leaves. The best-known chemicals are atrazine and simazine, which were developed in the 1950s following the discovery of their herbicidal properties by the Geigy Company in Switzerland. Related chemicals with slight modifications to the original structure include prometryn, desmetryne, terbuthylazine, and cyanazine. These are important pre-emergence herbicides in field crops. They are sprayed onto the soil and kill weeds as they emerge through the soil into the high.

Triazoles. Amitrole was introduced as a herbicide in 1954. It is a non-selective herbicide used to keep follow land clear, and against perennial weeds between trees in orchards.

Bipyridyliums. These important herbicides have played their part in the farming revolution. Functioning only in sunlight, they desiccate soft tissues such as leaves and so kill the plants on to which they are sprayed. Total herbicides, they have made a dramatic impact on farming since they were developed in the 1960s. Farmers plough fields to kill weeds and prepare a seed bed. But having tractors haul heavy machinery many times across fields costs money and damages the soil. So these plant desiccants were a clever answer to several farming problems: they could be sprayed on an unploughed field, killing every growing weed. Not only was ploughing then unnecessary, but also the dried weeds formed a protective layer on the surface of the soil, reducing erosion and providing a soft and protective surface mulch into which the seeds for the next crop could be gently inserted by machine. This was the advent of chemical ploughing.

Another advantage of bipyridyliums is that when they fall on the land rather than the plants, the chemicals are immediately and tightly bound to the soil. Thus they cannot be taken up by the roots, but, most important in chemical ploughing, they cannot affect the planted seeds when they germinate.

The two most often used herbicides in this group are paraquat and diquat. They are based on combing two pyridine rings, which gives them their group name bipyridyl.

Other heterocyclics Relatives include the substituted uracils like bromacil, which contains bromine in the molecule; terbacil with chlorine; and the thiadiazine chemical, bentazon. With the exception of bentazon, which acts on the leaf surfaces it contacts, these are also soil-acting herbicides.

Mode of action. In one way or another, all the heterocyclics disrupt the photosynthetic activity of leaves, starving plants of their energy. The triazines block important reactions creating ATP and NADP within the photosynthetic process. The triazoles block the plant's manufacture of the carotenoid pigments which protect the vital chlorophyll pigments in the leaves from destruction by sunlight. Dying leaves turn yellow as their green chlorophyll pigments disappear. The bipyridyliums work slightly differently by a sinister method. They need both light and oxygen to work. Oxygen is released in the chloroplasts during the photosynthetic reactions. The oxygen reacts with the paraquat or diquat, producing hydrogen peroxide, when destroys the membranes around the vital chloroplasts and other working parts within the planet cells.

Health and environmental effects. Most of the groups have low acute toxicity to humans but may cause skin or eye irritations. Over the term there may be cancer risks from simazine and amitrole, and both simazine and atrazine can form cancer-causing nitrosamines in the stomach.

The bipyridyls are more toxic than the triazines. They affect skin and mucosal surfaces of the nose, throat, and other respiratory passages. They cause ulceration and inflammation, and delay the healing of cuts and wounds. Diquat, and being swallowed, damages the gut, and paraquat tends to damage the lungs and kidneys. There is no known antidote to this poisoning. Both have produced deformed embryos in rats.

Atrazine and simazine are among the most frequently found pesticide contaminants of water supplies in the UK. Atrazine may be leached through the soil into underground waters or be washed into streams and rivers, where it can last for several years. The Government has placed both on a special 'red list' of dangerous substances over which there should be tighter controls in supplies.

Paraquat and diquat are strongly adherent to soil particles and break down rapidly on contact with the soil. However, they have been shown to kill blue-green algae in the soil, and paraquat, in particular, is toxic to fish and mammals, including hares. Users are advised to keep all livestock out of treated fields for at least 24 hours.

ORGANOPHOSPHATES

When the environmental persistence of the organochlorines was realised, the less persistent organophosphates were seen as the new generation of safer chemicals. Their toxic pedigree came from research into nerve gases during the Second World War. That they worked well as insecticides was clear, but the early organophosphate chemicals like dimefox were also very toxic to humans and safer versions had to the developed before they could be widely used.

The group includes a large number insecticides, which were the first organophosphate developed; herbicides; and fungicides. They all contain phosphorus and an organic structure that can be straight chain or ring. The group illustrates the great variety of structures among pesticides, with some members fitting equally in other families like the heterocyclics or pyridines. Thus, there are chlorinated organophosphates like fenchlorophos and also heterocyclic N-compounds like menazon.

Insecticides Among the group are such well-known members as parathion, which is no longer approved for use in UK. It is a non-specific insecticide that has a high toxicity to mammals (an acute LD_{50} of 13mg/kg for rats) and destroys non-target insects like bees, and birds.

The introduction of chlorine atoms in this groups was found to reduce toxicity to mammals without affecting the insecticidal powers. A series of such formulations appeared and was found, like many of the organophosphates, to be effective against insects which had become resistant to DDT, Bromophos is a contact insecticide against mites as well as caterpillars. It contains both chlorine and bromine and could also be included among the halogenated hydrocarbons.

A series including sulphur atoms sprang from demeton-S-methyl and includes phorate, a very toxic insecticide used against aphids and flies in crops like carrots and parsnips, and malathion. Malathion is an important treatment for aphids, thrips, and mites on a range of vegetables and fruits. It has a very wide range of susceptible insects and a low toxicity to mammals (an acute LD_{50} of 2,800 mg/kg for rats). It is a common ingredient in garden greenfly sprays.

Dichlorvos vaporises easily. It is used in household fly strips and has been used for flea collars. Its relatives, chlorfenvinphos and tetrachlorvinphos, fenchlorophos and mevinphos, are also non-specific insecticides of varying toxicity, all of which act on contact in the insect's stomach. Dichlorvos and chlorfenvinphos are highly toxic (with acute LD_{50} of 80 mg/kg and 15 mg/kg for rats, respectively) tempered to tetrachlorvinphos, which has an acute LD_{50} of 4000–5000 mg/kg.

Dimethoate is a systemic insecticide and acaricide used against mites, thrips, and aphids on food crops. It is more residual on sprayed crops than dichlorvos. Livestock should be kept out of sprayed fields for a week. Most sprayed crops cannot be harvested for 7 days, but watercress must be left for 40 days.

Another group of organophosphates is based on heterocyclic structures like the triazines and pyrimidines. Azimphos-methyl and diazinon are broad-spectrum insecticides. Chloropyrifos is a contact and stomach-acting insecticide with a very wide range of victims and a persistence of 2–3 months. Although these three chemicals are widely used in the home and garden, all are dangerous if swallowed and can irritate the skin.

Fungicides and herbicides Following the development of organophosphate insecticidal compounds, organophosphate fungicides were explored. Wepsyn was an early systemic fungicide based on the triazole structure. Systemic fungicides were a great advantage against fungi that penetrate and grow within the host plant's tissues. They could perfuse all tissues of the plant and kill any growing fungal threads which appeared. Pyrazophos, an organophosphate based on a heterocyclic nitrogen structure, is an interesting combination of systemic fungicide and insecticide.

A well-known and widely used herbicide is glyphosate, which is a broad-spectrum, translocated herbicide active against couch grass as well as other perennial weeds and annuals.

Mode of action. The mode of action of organophosphorus insecticides is the same anti-cholinesterase activity seen in the carbamates. However, different insects and higher animals handle these pesticides differently in the body, sometimes converting them to harmless products. This is why certain pesticides can be highly toxic to a small group of insects but not to others, or to humans.

The fungicides appear to act by disrupting enzymes, especially those associated with building the cells walls.

Health and environmental effects. The immediate effects of poisoning are seen in the functioning of the entire nervous system. First signs may be weakness and tightness of the chest with constricted pupils. Twitching and general muscle weakness may begin, and end in convulsions and respiratory failure. Very mild poisoning can lead to symptoms often mistaken for a virus infection, with sweating, headache, gastrointestinal upset, and weakness in the limbs. Such poisoning is difficult to diagnose because it is usually difficult to identify a source of low-level contamination with pesticides if it comes in food, grass, or dust. Low-level, long-term contamination can lead to damage to the nervous system, which may appear as depression, loss of memory, and anxiety. The organophosphates are easily absorbed through the skin, and severe irritation may occur where pesticides make contact.

Most organophosphates do not persist long in the environment. Exceptions such as glyphosate and chlorfenvinphos are strongly bound to soil particles, where they break down over a year or more. The group is very toxic to fish, birds, earthworms, and bees.

CHLORINATED PHENOXY SUBSTANCES

All plants have hormones that regulate growth. Natural plant hormones include auxins, gibberellins, and kinins. The auxins are concentrated in the growing shoots and are responsible for the elongation of cells, the bending of stems towards the light, and the downward growth of the roots. An excess of growth hormone can produce chaotic growth often seen as bulbous galls on leaves and stems of plants. This effects was turned into herbicide activity through the discovery of chemicals that mimic the auxins.

The first such herbicide was 2,4-dichlorophenoxyacetic acid, better known as 2,4-D. This had its greatest effect on broadleaf weeds, less on grasses, and least on woody shrubs and trees. This meant that 2,4-D and its family were effective as selective herbicides. A more widely used relative is MCpA (2-methyl-4-chlorophenoxyacetic acid), which is an ingredient in many commercial mixtures of herbicides. The controversial shrub and tree herbicide 2,4,5-T contains the deadly chemical dioxin as a manufacturing contaminant.

Other herbicides in the family include mecoprop and dichloroprop, which are derivatives of propionic acid. MCPB and 2,4-DB are closely related chemical based on phenoxybutyric acid.

Mode of action. The simulated hormone action of the family causes a disordered growth of the growing shoots which destroys any natural order or pattern of growth, and the plant dies.

Health and environment effects. These pesticides cause mild-to-severe skin problems, ranging from irritations to chloracne, which may be caused by the dioxin. The chemicals can irritate the lungs and gut, and affect the nervous system. Symptoms of nausea, vomiting, diarrhoea, and mental confusion are reported. There is evidence that 2,4,5-T is responsible for birth defects and cancers among agricultural workers, and it has been the target of a trades-union campaign to have it banned in the UK. Since the Vietnam War, in which both 2,4-D and 2,4,5-T were used to defoliate the forests, significant numbers of birth defects have been attributed to these herbicides.

The family is relatively harmless, to wildlife, specially as the chemicals are broken down rapidly in the soil. The dioxin contaminant persists longer than the herbicide, and it can affect reproduction is mammals. Evidence of a longer life in water comes from a Friends of the Earth study of contaminated drinking water in England and Wales that found mecoprop, dichloroprop, MCPA, MCPB, 2,4-DB, 2,4-D, and 2,4,5-T in both underground and surface water sources at levels exceeding the EC's maximum acceptable concentrations (MACs)

AMIDES AND UREAS

A large family of herbicides was developed out of urea-based, phenyl-ring-structured chemicals. Fenuron and diuron were early examples. More recent additions to the substantial list of field crop herbicides are monolinuron and isoproturon.

Related amine chemicals were developed, e.g. propanil and solan, which are not approved for use in the UK, and alachlor, which is. More recent versions are naphthylacetamide and pendimethalin. An example of the growing complexity of synthetic pesticides and the way the families become combined is the herbicide cyanazine. Its formal chemical name makes the point: 2-chloro-4-(1-cyano-1,methylamino)-6-ethylamino-1,3,5-traizine. It has chlorine, a heterocyclic ring with nitrogen atoms, amine groups and a triazine structure.

Another related group within the family are the anilines the provide herbicides like trifluralin, which is widely used in herbicide mixes.

Mode of action. The ureas disrupt photosynthesis and starve the plant of energy. The addition of chlorine atoms increased this effect but also increases the persistence in the soil. The anilines block seed germination and root growth by the disrupting mitosis, or cell division.

Health and environmental effects. Most are of moderate-to-law acute toxicity but, because of the nitrogen component in the urea and amide structures, they can produce cancer causing residues in the body. They are persistent in the environment. Many are used as residual herbicides to kill weeds emerging after the crop is planted. They are dangerous to fish, diuron especially so, and to algae in rivers or lakes which they reach in runoff.

BENZONITRILES

This family of herbicides is important in field-crop weed control in Europe, usually in mixtures with other herbicides. They are contact herbicides against broadleaf weeds. The two most popular members each contain a halogen atom — bromine or iodine — which is reflected in their names. Bromoxynil and ioxynil are popular in agriculture, and illustrating the tendency towards mixing active ingredients in commercial products are the 25 different mixtures using bromoxynil and the 28 using ioxynil approved in the UK. Many of the available mixtures contain both chemicals. Neither of them is allowed in household herbicides in the UK.

Related herbicides are dichlobenil and its cousin chlorothiamid, in which sulphur replaces nitrogen in the original nitrile structure.

Mode of action. They disrupt the energy storage and transfer system in plant, and they also disrupt photosynthesis.

Health and environmental effects. These chemicals are harmful if swallowed or if they contact the skin or eyes. They block cellular energy-transfer processes in humans and also produce symptoms of increased cellular metabolism evident as fatigue, sweating, insomnia, and, over time, loss of weight. There is no known effective treatment, but symptoms can be relieved.

Benzonitriles are designed to persist in soil and on treated plants. It is advised that livestock be kept out of treated fields for at least six weeks. They are also harmful to fish, which is important because they persist on the foliage of their target plants, which may find their way into nearby ponds and rivers.

PHENOLIC COMPOUNDS

The basic phenolic structure is a benzene ring with one hydroxyl group (OH) attached. The phenol of group of pesticides has members with additional atoms or molecules attached to the ring at one or more of the ring's carbon atoms. The group contains herbicides and fungicides.

It has long been known that phenols are useful against human infections caused by bacteria, fungi, and intestinal worms. Plant medicines like thyme and garlic owe their antiseptic properties to phenols, as does the common weed, tansy, which kills human infestations of thread-worms. Even the biblical myrrh was used as a fumigant, thanks to phenol. Phenols were used as early fungicides in agriculture but were often toxic to the plants they were meant to protect. Such compounds were suitable for 'dead' plant materials like sawn timber. Creosote oil is often used to treat wooden fencing in the UK.

The toxicity of these chemicals is reflected in the fate of the major dinitrophenols under the UK approvals systems. From creosote came 2,4-dinitro-o-creosol, or DNOC. It is a strong stomach poison against many insects, also killing their eggs, and also mites, but is also used as a contact herbicide against broadleaf weeds. Its approval for use in the UK was withdrawn at the end of 1989. Its close relatives, dinoseb and binapacryl, have had their approvals withdrawn since 1987. Dinocap is a powdery-mildew fungicide now in use but under review.

Sodium pentachlorophenoxide is a timber and wall fungicide and mossicide used in mushroom houses in the UK. It is related to pentachlorophenol, which was discussed under the section on organo-chlorines.

Mode of action. These substituted phenols uncouple oxidative phosphorylation, which is the means of storing and transferring energy in plant cells. The loss of the transfer ability means that the organells within the cells soon run out of energy supply, and the cells stop functioning and die.

Health and environmental effects. These phenols are toxic to humans and other mammals. They are absorbed through the skin and gut, and can damage the liver, kidneys, and brain. In the environment they are very toxic to fish and algae. They can vaporise from treated woods and are readily leached into surface and ground water supplies.

SYNTHETIC PYRETHRUM COMPOUNDS (PYRETHROIDS)

Pyrethrum, from the flowers of *Chrysanthemum cinerariaefolium*, is one of the oldest natural insecticides. It is also one of the few natural compounds still widely used, and is enjoying particular popularity today with the recent increase in environmental awareness. Pyrethrum has a rapid 'knockdown' effect but some insects may recover and fly on to annoy another day. It is usually mixed with other, usually synthetic, pesticides to kill the insects while they are stunned. It is very safe for other wildlife and has very low persistence. In fact, contact with light and air begins its breakdown. This means that it is poor replacement for some of the less safe pesticides now used in agricultural since most farmers want to pesticide to remain active for a long as it necessary to protect a crop or animal.

The commercial pyrethrum insecticide is made of a collection of six main chemicals from the flowers, each of which contributes to the insecticidal activity. The group is known as pyrethrins and individually as pyrethrin I and II, and jasmolin I and II. However, the names *pyrethrum* or *pyrethrin* on the product label should not be taken as a mark of safety. Many products are mixtures including some of the more toxic pesticides like lindane.

The pyrethrins have been synthesized, and many modified formulae are available, including isomers (compounds with the same formula but having some atoms arranged in a different pattern). The modified forms are known as pyrethroids. They include allethrin and its isomer, bio-allethrin; resmethrin and its isomer, bioresmethrin; and permethrin. Again, these compounds are often found in mixtures, sometimes with other pyrethroids or with pesticides like dichlorvos or lindane. The pyrethrins are often mixed with another chemicals known as synergist that acts to increase the activity of the pyrethrin. Piperonyl butoxide is often used, and it makes the mixture seven times as affective.

Residual pyrethroids have been developed, like cypermethrin and deltamethrin, that are less sensitive to environmental break down and are stronger contact and stomach insecticides than the earlier derivatives of pyrethrum.

Mode of action. The whole family acts by interfering with the transmission of nerve signals in both the brain and the nerves among muscles and other organs. Small doses causes repetitive and erratic nerve signals, but higher doses stop all transmissions of signals.

Health and environmental effects. The natural pyrethrum is relatively safe to mammals, including humans, but the synthetic pyrethroids have varying toxicity, and they should be treated with the same respect as other pesticides. The more recent member like deltamethrin are more toxic to humans and cause skin and eye irritations. Some members like permethrin and allethrin are seen as possible mutagens or carcinogens.

Because the pyrethroids are not selective, they kill non-target insects, and they are particularly dangerous to bees. The later, more residual versions are a particular problem. Deltamethrin, for example, persists for three to four weeks on sprayed crops. All pyrethroids are extremely toxic to fish.

PESTICIDES AND THE ENVIRONMENT

Agriculture is the main destination for chemicals sold, with 83% of all pesticides endings up on farms. Of all different type of pesticides, herbicides are the most common and once again the heaviest user is the agriculture industry. It account for 54% of all pesticide sales.

However, pesticides are also used widely outside agriculture in ways that can have significant for less than 20% of the total used, but they are no less important. The risks to user, consumer and to the environment remain and may be higher depending on the specific action required and the chemicals chosen. Both where and how pesticides are applied can be as important as specific chemicals ingredients in causing unwanted effects.

Although the greatest proportion of the three main groups of pesticides goes to agriculture, significant amounts end up in household or garden use. Less than 1% of the fungicides produced enter the home as pesticide, but 5.2% of all herbicides go on gardens, and 26.6% of all insecticides are sprayed in the home. Household pesticides are a good example of how little we know about the pesticides around us daily. Most home users pay too little attention to what is sprinkled or sprayed about their houses despite having to breathe sprayed air and eat off, or live on, pesticide-treated surfaces. But this is not the users' fault:

- Most pesticides which are approved for home or garden use are the less harmful end of the range, but this does not mean they are harmless.
- Household product labels convey far less information on both content and safe application than those on products

for industrial use.

- Unlike the rules facing industrial users, there are no rules or regulations governing how consumers can apply pesticides at home.
- The dangers of this are made worse by the total lack of training available for consumers in safe pesticide storage, handling, and use.
- The proximity of children food, and pets to pesticides used in the home and garden highlights the dangers. Most pesticides used at home are sprayed in the same environment where children and pets play, eat, and sleep. Avoiding contact during or after spraying is often impossible.
- The marketing of pesticides to industrial users now places great emphasis on images and messages of safety, but the risks are played down in presenting pesticides to household users, have you eve seen an advertisement for fly spray with the use in a gas mask and protective clothing? Have you seen an advertisement for pet flea-collars warning that pets should not be cuddled while warning the carbaryl-, diazinon- or lindane-impregnated collars for fear of contamination?
- Home users are given no idea of how little pesticide to use for effective and safe control. With pre-packed aerosol, sprays, they also have no idea how much chemical is emitted in a burst of spraying. Most home users sprays excessive amounts, which wastes money and greatly increases the risk of harm to members of the household. Farmers for examples, not only have guidance on optimum application rates, but also have to keep the cost of chemicals in the mind and thus are less likely to overspray crops.
- Contractors treating woodworm or dry rot, and Local Authority vermin control officers who spray for wasps and fleas or leave baits for mice and rats, too often fail to fully inform householders of safety precautions which will reduce their chances of being contaminated with the pesticides used.

Outside the home, manufacturing industries and public utilities are major pesticide users, though consumers are normally unaware of the presence of pesticides in the goods or services they supply.

NON-AGRICULTURAL USERS OF PESTICIDES

Household

- Household uses include dichlorvos **vapours strips** and **flea spray**; carbaryl and diazinon in pets' **flea collars**; pyrethrum-based **aerosol insecticides**; and lindane in **anti-and moth-sprays**.
- **Garden insecticides** contain lindane, dichlorvos, malathion, and are often mixed with fungicides like benomyl, captan, and triforine. The battery of herbicides includes 2,4-D, glyphosate, MCPA, and amitrole. Metaldehyde is in **slug baits**, while warfarin and difenacoum replace the Pied Piper in the **rat and mouse control**.
- The **woodwork** and treasured **wooden furniture** may be protected from fungal or woodworm attack with chemicals like lindane and tributylin oxide, while the emulsion paints on the paintwork could contain pentachlorophenols to keep fungi off the bright pastels.
- The **bricks and mortar** supporting the treated woodwork may be washed with antifungals like dodecylamine salicylate.
- **Wallpapers** and their adhesives contain slimicides to block moulds. The chemicals enter the production in the 70 tonnes of water used in making 1 tonne of paper. A large proportion remains in the paper, and the test goes into the environment. Pentachlorophenols, dichlorophen, and benzothiazoles are also approved for use.
- **Carpets** are often treated with moth-proofing chemicals, including common insecticides like pyrethrums and lindane.
- **Tap water** contains residues of a large number of pesticides from agricultural and industrial users as well as a smaller amount from pyrethroid insecticides, which are

deliberately added to control the water hog-louse in the mains pipes.

Manufacturing Industry

Among the vast range of manufactured goods are many with significant amounts of perishable components, to which pesticides are applied.

- The **paper** industry has been mentioned already.
- **Natural-fibre textile** manufacturers use antifungals, of which pentachlorophenyl laurate used to be especially. The fungicide not only has to kill and be persistent, but must also adhere to the textile to retain efficiency. The antifungals penta-and tetrachlorophenol have been used in the manufacture of textiles from cotton, flax, and jute, and with products like fabric, cordage, and tentage. Copper and zinc naphthentes have been effective on rope and sacking.
- **Important leather** or leather goods may be treated with chemicals like naphthenates, lindane, dieldrin, sodium arsenite, and even DDT, although this is not permitted on UK-made leathers.
- **Adhesives** for wood, paper, leather, and other surfaces are often based on cellulose, starch, and milk protein, or casein, and they are protected with fungicides and insecticides, including pentachlorophenol, tributylin oxide, sodium metabisulphite, and borax. Even the water-emulsion adhesives based on polyvinyl acetate and polyvinyl alcohols are susceptible to attack and contain pesticides.

THE ENVIRONMENTAL ASSAULT

Environmental damage from pesticides is dependent on two main factors: how pesticides are used: and how they behave once let loose in the environment.

Most pesticides are blunt weapons. Using them is like inviting a motorised road-sweeper into the dining room to collect the crumbs from under the table, or amputating a foot to remove the pain of a bunion. Both may work out at the undeniable cost of significant additional damage on the way.

Pesticides kill organisms other than the target pest. It is necessary only to read a set of data sheets for internationally available pesticides to see the potential scything of organisms. Whether for fungicides, insecticides, or herbicides, the reports of toxicity losts often record significant toxicity to one or more of the following bees, quail, fish, mammals dogs, mallard ducks, etc. Those organisms are mentioned only because they happen to be used in the toxicity-testing programmes required for the approval and licensing of new chemicals. There is no similar record of the *total* range of species affected because such observation are rarely, if ever, made. Bats, for instance, are particularly vulnerable to lindane and dieldrin, and the use of these pesticides as a treatment for roofing timbers is thought to be responsible for the decimation of UK bat population. Lindane is still used in these products.

Insecticides readily decimate the local insect population on sprayed crops, taking along only the natural predators of the target pest, but also harmless, and even beneficial, insects. Honeybees get clobbered when orchards and other flowering crops are sprayed. But they are also at risk when aphids are sprayed on crops no longer in flower. Bees gather the sugary secretions from the tails of feeding aphids and can either pick up spray aimed at the aphids or get caught in the spraying process. It matter little whether toxic insecticide or herbicide is being sprayed. Strong odour from pesticides can mean bees returning to the hive being killed by guard bees who identify them as foreigners. The British Agrochemicals Association (BAA) reported a minimum of 353 bee colonies affected by pesticides during 1988 in the UK.

Fish have an unfortunate skill in extracting the tiniest amounts of pesticides from water through their gills as they take their oxygen supplies. In one mishap in Canada, 70,000 trout died when herbicides were sprayed along a roadside ditch and ran off into a trout farmer's water supply. Toxicity tests repeatedly show the sensitivity of fish to the complete range of pesticides. Few chemicals are sprayed directly onto water. Runoff from fields, contamination of drainage ditches when spraying weeds one the banks, seepage, through soil, and irresponsible disposal of empty pesticide containers are efficient ways of contaminating water. It is estimated that 88% of modern pesticides affect fish.

One fish, salmon, is treated with the pesticide Nuvan 500 or Aquagard (dichlorvos) to kill the sea lice that grow on farmed fish. While the salmon may benefit from the loss of lice, there is concern that crustaceans and other shellfish living on the floor of the lochs may be suffering. So recent is the boom in salmon farming in Scotland that there is insufficient research data to say how serious damage to other life in the lochs may be.

Plants and other wildlife around sprayed fields can be affected by crop sprays. Both herbicides and insecticides are often applied in mixtures, with several ingredients of each type of pesticide giving a wide kill-range. This has disturbed the ecology around fields, especially the large fields of East Anglia where hedgerows have been grubbed out, and the headlands dividing fields are narrow refuges from the plough-and-spray boom. Butterflies decline through the combined effects of direct attack from pesticides and the reductions of natural feed plants for the larval stages.

Many new agricultural sprays use their broad spectrum of target pests as a selling point. The herbicide Invicta (bifenox and isoproturon) is promoted as giving 'super league performance . . . It's outstanding against Black-grass and over 40 other weeds in winter cereals.' Another cereal herbicide is called Harmony M (thifensulfuron methyl and metsulfuron methyl), demonstrating how loosely the English language can be used in the service of marketing. It boasts of the 'broadest spectrum of weed control available for winter wheat. On its own it takes out over-wintered Mayweed and Chickweed, Speedwell, panes, and Forget-me-not, as well as spring germinators like Knotgrass, Redshank, and Black-bindweed . . . Quite simply it's the broadest spectrum control in its field.' But there is evidence that such total control of weeds may not always offer an advantage to the farmer. Work with rape in Scotland and on the Long Ashton Research Station in England showed that there was no loss of yield with as much as 30-35% of the ground covered with broad-leaved weeds. As the rape crop grows, it shades out the lower growing weeds.

The soil, which feeds and physically supports food crops, is not mere 'dirt' but another world rich in living organisms vital to the survival of life above ground. Soil has complex interconnections between its physicals and chemicals components which provide structure, the 'nutrient glue' humus and a rich population of organisms. There are bacteria in numbers approaching 3 billion per acre or, for those readers more visual than numerate, around 1.5 tonnes of live bacteria per acre. There are also actinomycetes, close relatives of fungi, and large numbers of fungi, especially in rotting leaves or other vegetation. These produce wispy white mycelia through the soil and occasionally send surface-wise their fruiting bodies known as mushrooms and toadstools. There are abundant algae, and mobile organisms called protozoa, which are single-celled, animals including the amoeba. Larger and visible to the naked eye are the mites, springtails, earthworms, slugs, ants and termites. In all, there might be around 25 million of these larger creatures in each acre of land, including 1.5 million earthworms.

The combined activities of this population include: producing nitrogen and carbon dioxide for higher plants; breaking down for reuse the vast quantities of plant and animals debris falling onto the soils each year; and keeping the soil structure suitable for air and water to move freely. Most of these organisms live in the top six inches of soil. Earthworms in particular are killed by pesticides like carbofuran, which aims at soil nematodes, and the leaf fungicide benomyl. Benomyl also kills beneficial soil fungi, which, in turn, are vital food for a predator of aphids. Cereal fungicides applied to protect crop leaves have been shown to damage the soil organisms and cause additional pest problems for the crops. The fungicide washes into the food soil and kills soil fungi vital to the local insect population's food chain. A key predator of the aphid is thus starved to oblivion, and the aphid population explodes, causing farmers to apply additional insecticides and fungicides.

Insurance spraying

Many pesticides are applied in far greater numbers and quantities than are necessary to achieve appropriate control of pests. A large-scale cereal farmers not only has a large potential return in 1,000 hectares of waving grain, but also a large bill for each passage over those hectares to spray weeds, insects or fungi. There is an understandable preference for 'insurance' spraying in the hope of the greatest number of likely pests being attacked for the longest time with the smallest number of spraying. This means more chemicals, a wider range of target pests and often higher-than-advised doses, just to be sure. Cosmetic sprays on perishable crops invite unnecessary spraying. Fruit and vegetable growers face demands for 'blemish-free' produce and risk rejection of an entire harvest if blemish tolerance are exceeded. Whether using dimethoate on lettuce on potatoes, growers are compelled by the market to apply 'insurance' pesticides even when there may be evidence of the pests or only slight damage. From the hyperbole of sales reps to the regulations on the pesticide use, there is little to discourage such dangerous profligacy. Hope for the future lies in the development of new approaches, especially Integrated Pest Management (IPM).

Climbing the food chain

Not only can pesticides affect the food chain at any point through accidental contamination but they can also spread up through the food chain in multiplier effect, killing organisms unrelated to the targets. Birds and mammals eat poisoned insects or smaller animals that may have eaten other animals that have eaten the poisoned insects. Thus each step up the chain finds higher organisms eating large numbers of organisms below them and, as a result, accumulating increasing concentrations of pesticides in their body tissues. Birds of pray are most susceptible. During the 1970s, there were only 10 pairs of peregrine falcon in the hole of Canada between the Rockies and the Atlantic coast. In the UK, similar effects have been noted on peregrine falcons and kestrels among the carnivorous birds, but also on pheasants. The Royal Society for the Protection of Birds (RSPB) has recorded bird deaths from poisoning as rising from 58 in 1982 to 103 in 1988. Incomplete figures for 1989 are already 157. Not all of these deaths are due to the multiplier effect. Some are caused by pesticides used in basis laid deliberately to poison the birds, usually birds of prey. Gamekeepers breading groups or pheasants for shooting are often found responsible. In 1987, the toll included 2 golden eagles, 1 peregrine falcon, 12 buzzards, 6 rooks, and 17 crows.

At the top of the food chain are humans, and we are as vulnerable as other animals to off-target poisoning. There is evidence that our health is affected by residues in food as well as by direct contact with pesticides in their manufacture or application. The general background contamination by pesticides undermines our health no less than that of wild pansies or falcons. persistent chemicals DDT are found in fat tissue of all humans tested, even in newborn babies. Human breast milk might be seen as the very top of the food chain. The level DDT in American farm workers was found to be 23 ppm (parts per million) in 1972, and the average US citizen had 8 ppm. Just for a perspective on that level of contamination: were the flesh of Americans to be sold of imaginary cannibals in the EC, the meat would be declared unfit for consumption, as it contain 8-23 times the EC Maximum Residue Level (MRL) of DDT, UK levels of DDT in human fat may have fallen recently, but government figures show that in 1977, both men and women still had DDT levels in their fat that exceeded present ECMRL-an average of 2.6 ppm for men and 1.6 for women. Some men had 15 times the legal limit and some women 17 times. Not even EC pigs a allowed to carry the amounts of DDT in body fat tolerated by humans in the UK.

Pesticides Persist

The ideal pesticide would kill its specific target and then disappear without trace. Sadly, this rarely happens. Many chemicals are designed to last long enough to kill posts that are expected to make an appearance *after* spraying. When lindane is applied to household timbers, it is meant to last for years. In farming, many herbicides and fungicides are applied in the hope of lasting through most of a crop cycle -5 to 12 weeks- or over the winter months before active growth begins in the spring. Many of these pesticides accumulate in the soil and kill soil flora and fauna. The more application -especially if the pesticides are held on clay or humus in the soils- the more pesticides build up, and the more damage they cause.

Many pesticides last longer than necessary to do their job. DDT is the best-known example, persisting for approximately 10 years in the environment. Lindane, or BHC, can last 14 years , and common herbicides like paraquat and simazine last 3-5 years. There is usually a choice though, with malathion, for instance, braking down in days, which explains its recommendation for use on household vegetables.

Nomadic Pesticides

Wandering pesticides can cause havoc. Few become immobilised, like the herbicide atrazine, on contact with the soil. Many travel long distances, invisibly, appearing not only where least expected but where they may never have been applied deliberately. The longer a pesticide survives, the further it may travel.

The water route

Water often facilitates wandering. Runoff from field, seepage into underground watercourses, accidential spillage or deliberate dumping into rives and drains, are the main routes. Sometimes pesticides are added deliberately to water. Pyrethroids, for example, are used to kill the water hog-louse in mains pipes. The chemical can be added at levels as high as 10 micrograms per liter (mg/1) for up to 7 days continuously. This levels is 100 times the EC legal limit for a pesticide in drinking water. Contamination of underground water may be more severe in lighter soils. In East Angila the water authority has reported extensive contamination by agriculture pesticides.

According to a 1987 FoE survey of pollutants in drinking water, at least 298 water supplies in England and Wales exceeded the EC's Maximum Admissible Concentrations (MAC) for a single pesticide (0.1 μg/1), and 76 exceeded the MAC for total pesticides (0.5 μg/1). Seventeen different pesticides were found above the MAC, but this was probably an underestimate because the data came form the water authorities, who do not routinely test for all possible pesticides. The supplies exceeding the MAC were clustered in the South East-East Anglia, the Home Counties, London and Wiltshire. The UK government is breaching agreed EC law in failing to reduce these levels.

Problems associated with contaminated underground resources include the greater time pollutants take to 'wash out', and apparent delayed breakdown of the pesticides. Among pesticides found in three underground water sources at levels above the MAC was the banned DDT. Additional concern comes from the lack of data on both the breakdown products of pesticides and the new products formed as pesticides react with other chemicals in the water supply. Chlorine, which is used routinely in water routinely in water treatment, can react with herbicides like 2,4-D to produce chlorophenols, which are human carcinogens. The pesticide breaks down into aldicarb sulphoxide, which is more toxic than the aldicarb.

The atmospheric route

Pesticides also move in the atmosphere as spray drift from

ground or aerial applicators. Particles can drift thousands of meters, becoming more concentrated as their water or other solvents evaporate. The new Food and Environment Protection Act, allows and aerial spraying within 200 feet of houses, school, or hospitals, but requires a minimum of three-quarters of a mile between any nature reserve and spraying. Aerial spraying is allowed to continue in a 10-mph wind.

People in the vicinity of any sprayers can easily be exposed to drift, especially if they are down wind, however slight the wind. Pesticides are often applied in what is called Ultra-Low-Volume (VLV) applicators, which dispense a very fine mist, the particles of which can be carried for miles in the atmosphere. Other forms of application include fogging and smokes, which also produces ultra-small particles that can travel long distances.

The trade route

Industrial countries have developed sophisticated means of producing foods in one area and then shipping them to where they are eventually eaten. This movement of food involves the whole world. Ingredients like spices are herbs, fruits, and vegetables, and manufactured foods are shipped back and forth from continent to continent. Putting aside the gasronomic delights this allows, it is a highly efficient transport system for pesticides. Most pesticides are used on growing food crops or during storage and transport. Known politely as 'residues', these hitchhiking pesticides travel not only from British farm to dinner plate but significant amounts also appear on important foods in the UK and any other country.

Importing foods presents a particular danger. While one country may have strict regulations on which pesticides can be used on its home-grown crops there can be little control on what happens elsewhere. Pesticides banned or severely restricted by one government can be regularly on food and served up to the local residents. Despite the long-standing ban on the use of DDT in the UK, the government's Working Party on Pesticide Residues had to report in 1990 that imported pork from the Republic of China Regularly exceed the EC's maximum Residue Level (MRL).

Pesticides interact

It seems remarkable but is nevertheless true that modern

pesticide testing takes little account of the ways pesticides are mixed in practice. The mixture may occur either in a single application or through successive application of different mixture on one crop, in one space like the home.

Pesticides can also react with other chemicals in the environment, producing different and dangerous byproducts. The approved fungicide maneb degrades to the toxic ethylene thiourea (ETU) or ethylene thiuram disulphide, while several pesticides, including MCPA, 2,4-D, and benazolin, degrade to carcinogenic dichlorophenols. Pesticide products contain solvents, fillers and emulsifying agents as well as the active ingredients. These substances may also be highly toxic. However, little information is offered on the combined toxicity of the active ingredients plus the 'carrier' chemicals when products are offered for approval, and less is available on how they react once in the environment. The insecticide heptachlor is converted by microorganisms in soils to a more toxic and persistent chemical than the original. Even the persistent organochlorine DDT is changed in plants to two breakdown products, DDD and DDE, or in animals to TDE and DDE. Insects that developed a resistance to DDT often did so by knocking a chlorine atom off the DDT molecule to form DDE, which was harmless to them.

Synergism is another trick performed by some chemicals. When two synergistic pesticides are used together, their combined effect is greater than the sum of the individual effects. Piperonyl butoxide is added to pyrethrum insecticides as a synergist. DDT activity is increased in resistant insect population if a chemical called WARF Anti-resistant is added. This blocks the enzymes in insects which can convert toxic DDT into safer-for insects anyway - DDE.

However, accidential synergism can occur with pesticide mixtures. Apparently 'safe' amounts of fenitrothion, an organophosphorus insecticide, appear to react with equally small amounts of DDT, causing disproportionate live damage, followed by brain damage which mimics a well-documented medical condition called Reye's syndrome. The danger of synergism is that both its occurrence and its effects are unpredictable. Added to our ignorance of how any *single* pesticide may behave in the environment and our ignorance about the interactions between pesticides, the additional complications of synergism magnify our vulnerability.

Pesticides Can Encourage Pest Resistance

Nature is rarely stupid. As toxic chemicals are poured onto organisms the organisms often change to produce resistant offspring through mutations. This can render future applications of the pesticide useless, may tempt the use of higher and higher doses in the hope of success, and sometimes creates a worse pest problem. Environmental damage can be increased by the common response of producing commercial products with a wider spectrum of target organisms.

The pesticides rarely causes the mutation; rather, it helps speed up the selection process by removing from the population the susceptible pests and leaving only those with the newly inherited resistance. These new Adonises are left to multiply, thereby spreading the new resistance through the population. Often it is a simple mutation change, creating new enzymes which can either break down or detoxify the pesticide. But there also impressively inventive results of mutation that leave one in unashamed awe of the machinery of evolution. One of the earliest insects to develop resistance to DDT was the malaria-carrying mosquito. Some resistant species beat the worldwide eradication programmes simply by refusing to enter houses where DDT had been sprayed. In another example of resistance, the young larvae of the apple codling moth evolved the knack of spitting out the first bit of fruit treated with stomach-poisoning insecticides.

Resistance is seen in all pest types. In 1955 there were only 50 species of insect recorded as resistant to pesticides. This increased to 185 during the next 10 years and , by 1975, there were 305 species known to be resistant to available pesticides. Given that the range and sophistication of pesticides was increasing over this period, resistance was winning. Weeds have shown stubborn resistance to the s-triazine herbicides, which are the most widely used on field crops. By 1981, there were 29 species resistant to this group; 37, only 2 years later; and 42, by 1987. The impact and scale of resistance is evident in Hungary where 75% of all agricultural land was infested with a triazine-resistant weed, *Amaranathus* (a close relative of love-lies-bleeding), as early as 1975. Weed resistance to paraquat, diclofop-methyl, and trifluralin herbicides is also widespread. In the UK, the four main broad-leaved weeds in cereals are still a pest despite 20 years' intensive use of herbicides and the apparent improvements which manufacturers boast of. Hop growers and nursery-stock growers are landed with herbicide-resistant annual meadow grass.

Fungi, which are simpler organisms and which reproduce more rapidly than higher plants or insects, develop their resistance even more quickly. The more specific the action of the pesticide, the more easily resistance develops. In one experiment with a common mould, *Aspergillus nidulans*, nine strains of the mould were treated with ultraviolet radiation. Five strains mutated sufficiently to become resistant to the fungicide benomyl. Ultraviolet light is part of the sun's radiation and causes similar damage in humans which may result in skin cancer. In practice, commercial crops have shown the same speed of developing resistance. The first evidence of field resistance to benomyl was seen in the second year after it was launched, resistance to dimethirimol was shown in its first year, and resistance to the cereal fungicide enthirimol was shown in its third year.

The response to resistance has taken several routes. First, the agrochemical industry is trying to increase the rate of development of chemicals with different modes of action. Second, aware that the life of any new product is limited to the time taken for resistance to develop, the industry is trying to vary the mix of chemicals and the modes of action in their new commercial products. With no one chemical in use for more than a year or two and with deliberate variations in the modes of action, resistance cannot develop so quickly. Manufacturers also advise using different chemicals at different times during a season to try to slow down the rate of resistance development. Farmers are encouraged to follow manufacturers' advice strictly and to follow practices like choosing resistant varieties, rotating crops, and using strict crop hygiene and biological controls. Ironically, many of these practices are just what the agrochemicals were developed to reduce or replace.

PESTICIDES AND HUMAN HEALTH

There is no doubt that pesticides harm human health. The World Health Organization estimates that there are 3 million severe acute poisonings worldwide each year and 220,000 deaths attributable to pesticides. It is no great comfort that only 1% of these deaths occur in industrialized countries. The main difference between the deaths in less developed countries and countries like the UK lies not in safer pesticides but in greater care in using them. More careful use of a dangerous substance does not reduce its potential harm.

The dangers of pesticides are well explained by a scientist writing for users:

> It is necessary to bear in mind that all [pesticides] are biocides, and that at a subcellular level most organisms are dependent on similar chemical processes. Consequently, it is inevitable that most pesticides will continue to have adverse effects in foe and friend alike. Even those based on natural products or those directed at specific 'non-human' targets cannot be assumed to be free from risk since they are still biologically reactive and may interact with totally different targets in mammals.

There are two categories of harm that must be considered. First is the acute effects. These cover the immediate effects which appear soon after poisoning. They may be caused by a major exposure to pesticides—such as breathing fumes or spelling mix onto hands or face-which could hardly go unnoticed. Or they may be caused by a minor exposure—like brushing against recently sprayed crops-which can easily pass unnoticed. Such minor poisonings are not easily linked to their cause because people may have no idea of when or where they were in contact with pesticides.

The effects of major exposure may be serious enough to result in hospitalization or death, but they are just as likely to cause less dramatic effects such as rashes. The effects of minor exposure are usually less serious and include headaches, breathing difficulties and rashes.

Longer-term or chronic effects are the second category of harm. They are usually caused by continuous exposure to very small amounts of pesticide, often over years. Chronic effects include very serious diseases, like cancer, and also such minor complaints as skin rashes.

ACUTE POISONING

No one seems to know how many people are affected each year

by pesticides significant number of pesticide-linked deaths occur but are either not recognised as such or simply don't get recorded. Add to these omissions the dismal fact that the organizations don't pool their data — partly because they collect different data and use different criteria — and the national statistics on human harm from pesticides are inadequate, if not useless.

Illness, as opposed to deaths, from pesticide exposure are even less well recorded. Most minor exposures caused relatively minor acute symptoms. These include nausea, dizziness, vomiting, abdominal pain, diarrhoea, lassitude, headache, sweating, and thirst. Most people experience one or more of these symptoms often enough to dismiss them by saying 'I' am not feeling too good today.' Few people would go to the doctor with them and, if they did, few doctors would even guess at pesticide poisoning.

Certainly, if few cases of minor acute poisoning are seen, much less recognised by general practitioners, it is not wonder that the incidence of acute poisonings among the general population is thought to be low. Again, the recorded data vary widely.

CHRONIC POISONING

The chronic effects of pesticide poisoning are more difficult to attribute to pesticides simply because they appear long after initial exposure to the chemicals, and they can be due to very small amounts of chemical over many years. It is important to realize that few chronic effects are attributable to such obvious exposure as swallowing or skin contact. It is the very small amounts of residue, or contamination from droplets or vapour over months or years, that are of concern, amounts so small that they are rarely noticed. Often they are impossible to identify because they are odourless, tasteless, and colourless.

Exposure may come from food, ordinary household items like clothes, pot plants and pets or from household chemicals such as those in garden sprays, flea collars, rat poisons and wood preservatives. Other sources include factories which make or use pesticides, and many other places of work where pesticides happen to be present. Because the presence of chemical traces is usually unknown, it is difficult, if not impossible, to link pesticides to any subsequent symptoms. Even when a chemical may be suspected, proving the link is extremely difficult. Individuals will have been in contact with a large number of other substances during the time it takes for chronic effects to appear.

Mutagens

The most feared chronic effects are cancers, and the deformities to developing embryos in the womb. These defects can be caused by mutations which may be due to pesticides and other chemicals. Mutations caused by mutagens are permanent and inherited changes to genes, or to the chromosomes that contain the genes. The damage can be to a small segment of a chromosome or a gene, or can affect a larger length of a chromosome. Often chromosome broken by mutagens are rejoined by protective processes in the cell but are resembled incorrectly. Very small mutations can occur within a gene when cells are dividing and the chromosomes are making exact copies of each other. Such small chemical changes to genes may alter a single component part of a cell's protein, but that may be enough for it to be significantly different from the intended structure. If a changed protein is concerned with directing the behaviour of the dividing cells, the tissue may grow in chaos, because the genetic instructions telling the cell how to develop or when to stop growing have been severely disrupted. That is what happens in a cancer.

Few pesticides are deliberately formulated to cause chromosome abnormality, but insect chemosterilant are one example. They are often designed to interfere with the target insect's chromosomes, causing sterility. Other pesticides, however, may be transformed within the target organism, or other living things, and produce mutagenic by-products. The widely used herbicide atrazine has been shown not to be mutagenic in the pure state but to produce mutagenic products when taken up by maize plants. In essence, it is impossible to predict mutagenic behaviour of chemicals from their formulae alone. Mutagenic pesticides in common use include benomyl, captan, carbofuran, chlorfenvinphos, cyanazine, dichlorofluranid, dimethoate, omethoate, paraquat, simazine, and thiram.

Carcinogens

Cancers are chaotic cells that have lost their essential instructions. They occur in tissues that divide and reproduce in the body—blood-forming tissues, skin, intestines, and sperm—and ova-pro ducing organs. It is thought cancers need both an initiator and promoter of growth. A chemical may initiate an irreversible muta tions which is a potential cancer but which may lie dormant for years until it is promoted into active growth. Once promoted, it becomes a cancerous cell and grows out of control. Cancerous tissue has cells that are only loosely stuck together, so it is easy for some of these dividing cells to slip away and travel to some other part of the body where they can settle and continue growing. This is called metastasis and is how cancers spread, causing secondary growths. So far it is not known why an initiator is an initiator, nor what makes a promoter promote. Sometimes the same chemical promotes and initiates, but otherwise, different substances are required. Promoters, or whatever type, seem to require many years of insidious presence in order to act. This is why cancer may appear only many years after the first exposure to a carcinogen.

The whole process of cancer causation and growth is still shrouded in mystery, making it difficult to establish the possibility of a pesticide being carcinogenic. It is even harder to protect against them. Most carcinogens are thought to be mutagens also, but not all mutagens are carcinogens.

Many pesticides are called either potential or presumed carcinogens on the basis of laboratory studies and experience with human populations. These are often classified with the quaint name of 'suspected carcinogens'. It is very difficult to extrapolate findings in animal experiments to human, not least because the same tests often produce totally different effects in different species. Since it is still unethical to conduct controlled trials with carcinogens on human, there is no acceptable way to test the validity of using animals to identify human carcinogens. Common pesticides thought to be carcinogens include aldrin (UK approval withdrawn in 1989), benomyl, captafol, captan, 2,4,-D,lindane maneb, macozob, thiram, and zineb.

Teratogens

Teratogens cause deformities to the unborn foetus, producing miscarriages as well as what are termed congenital defects. While mutagens may be teratogenic, other teratogenic agents include chemicals which damage the foetus without any mutation occurring. Chemicals involved are varied and include some medicinal drugs like certain anti-cancer medications and thalidomide as well as pesticides. They seem to work by interfering with cellular metabolism in very active growth-centres like those in the rapidly growing embryo. There is evidence that embryo defects can also be caused by defective sperm which have been damaged by pesticides. This was first recorded in men working with an early organochlorine insecticide, Chlordecone. Sperm damage also manifested itself as birth defects in the wives of Vietnamese soldiers who were exposed to the massive doses of herbicides dumped on South Vietnam by the Americans. Although some foetal defects are believed to occur naturally, it is thought that chemicals in the environment may be the cause of significant but difficult-to-estimate proportion.

Table 7.1 : Possible mutagenic, carcinogenic, and teratogenic hazards among pesticides

Pesticide				Pesticide			
aldrin*	M	C	T	dichlofluanid	M		
allethrin	M			dichlorvos	M	C	T
amitrole	M	C	T	dicofol		C	
atrazine	M			dieldrin*	M	C	T
azinphos methyl	M	C		dienochlor	M		
benomyl	M	C	T	dimethoate	M		
binapicryl*	M			dinocap	M		
boric acid	M		T	dinoseb*			T
captafol	M	C	T	diquat			T
captan	M	C	T	disulfoton	M		
carbaryl	M	C	T	dodine	M		
carbendazium	M	C		endrin*	M	C	T
carbofuran	M	C		ethylene dichloride	M	C	
chloramben	M	C		ethylene oxide	M	C	T
chlordane	M	C	T	etridiazole	M		
chlorfenvinphos	M			fenitrothion	M		
chlorpicrin*	M			ferbam	M	C	
chlorothanonil		C		ioxynil			T
cyanazine	M			lindane		C	T
cycloheximide			T	malathion			T
2, 4-D		C	T	maleic hydrazide	M	C	
demeton-S-methyl	M			maneb		C	T
di-allate*	M	C		MCPA			T
diazon			T	methyl bromide	M	C	
dicamba	M			naled	M		

Contd..

Table 7.1 : (Contd.)

Pesticide				Pesticide			
nicotine	M	C	T	rotenone		C	
omethoate	M			simazine	M		
oxine copper*	M			sodium chlorate	M		
paraquat	M		T	sodium nitrate		C	
pentachlorophenol		C	T	2, 4, 5-T*	M	C	T
pentanochlor		C		tetrachlorvinphos		C	
permethrin		C		thiometon	M		
picloram		C		thiourea*		C	
piperonyl butoxide		C		thiram	M	C	T
pirimiphos-ethyl	M			trichlorophon	M	C	T
pirimiphosmethyl	M			trifluralin		C	
prometryn	M			vamidothion	M		
propachlor			T	zineb	M	C	T
propham		C		ziram	M	C	
pyrazophos	M						

M = Mutagen, C = Carcinogen, T = Teratogen
*Indicates pesticides no longer approved for use in the UK.

Common pesticides under suspicion include aldrin (UK approval withdrawn 1989), benomyl, captan, captafol, 2,4,5-T, dichlorvos, and diazinon.

Table lists some of the chronic hazards that are established or strongly suspected from studies throughout the world. So difficult it to establish firm proof that it is easy to oppose any listing of carcinogenic or teratonic chemicals simply with the objection that firm, easily, reproducible and scientifically 'proven' evidence has not been presented in most cases. The failure of a chemical to appear on a 'danger list' does not necessarily mean it safe. In 1978, NIOSH published a list of possible mutagenic or carcinogenic pesticides to which workers were exposed during pesticide manufacture, but were unable to assess 1300 pesticides because there were no data available. Given these difficulties and the health and hereditary risks, it is more than prudent to respect lists of 'potential' or 'possible' hazards until more conclusive evidence emerges.

Other Chronic Effects

There are other chronic effects that are less frightening but still unpleasant and unacceptable. Included are slight, but accumulating, damage to kidneys, liver, lungs, and heart, as well as a range of nervous disorders and allergies. Allergies and sensitivities to pesticides are the result of the body's immune system recognizing some unwanted chemical or foreign body and going into a protective overdrive. The chemical, or body in the case of pollen in hayfever, need not be harmful to health, but this does little to reduce the distress caused.

Pesticides may produce reactions like asthma and dermatitis of various types, and they may induce hypersensitivities that cause subsequent reactions to very small concentrations of the initiating chemicals. Once susceptible people have reacted to an acute contact with a pesticide, subsequent attacks can be brought on by the merest trace of the same chemical.

With pesticide residues a normal component of daily foods, many hypersensitive people experience symptoms like abdominal pain, migraines, and fits which could be precipitated by these residues. The first dose of pesticide could have been years before and may not have been noticed by the sufferer. The difficulty of identifying the cause is made worse by the widespread occurrence of pesticides and other chemicals in the environment and the difficulty in finding uncontaminated samples with which to test these sufferers. Of course, not knowing what residues may be on food also makes it difficult for hypersensitive people to avoid symptoms. Even organic food can have pesticide residues from background environmental contamination.

Lindane, which is used in wood treatments and also head lice shampoos, is believed to result in reactions which include epileptic fits, headaches, and also a serious blood disorder called aplastic anaemia, which may not appear until months after exposure.

The immune system may also be damaged as well as overstimulated by pesticides, as in allergies. The existence of AIDS is now well known, but little is known about how the immune system can be so severely damaged and of the extent to which this immune system suppression occurs in the population. Parathion, an organophosphorus insecticide, has been shown to suppress the immune response in hamsters after a single dose. The effect is a general lowering of resistance to common infections and could be occurring in human populations.

Delayed damaged to the nervous system, so-called neurotoxicity, can follow exposure to pesticides. One of the main characteristics of long-term damage is destruction of the insulating myelin coat that surrounds nerve fibres. Organophosphorus pesticides that work by disrupting the sending of nerve signals seem to cause this long-term damage to the myelin coating. The results are muscle weakness and paralysis. There is also evidence of chronic disruption of the acetylcholine mechanisms for carrying nerve signals from nerve fibre to nerve fibre, from the brain to the body.

REFERENCES

Barthalmu, G.T. Terrestrial organisms, in *Introduction to Environmental Toxicology.* F.E. Guthrie and J.J. Perry, Eds. Elsevier/North Holland, Amsterdam, 1980.

Carsel, R.F. and Smith, C.N. Impact of pesticides on ground water contamination, in *Silent Spring Revisited*, G.J., Macro, R.M. Hollingworth, and W, Durham, Eds. American Chemical Society. Washington, DC, 1987.

Carson, R. *Silent Spring*, Houghton Mifflin, Boston, 1962.

Davies, J.E. and Doon, R. Human health effects of pesticides, in *Silent Spring Revisited.* G.J. Macro, R.M. Hollingworth, and W. Durham, Eds. American Chemicals Society, Washington, DC, 1987.

Duffus, J.H. *Environment Toxicology.* Edward Arnold Ltd., London, 1980.

Edwards, C.A. *Persistent Pesticides in the Environment*, 2nd, ed., CRC Press, Cleveland, OH, 1973.

Freed, V.H. Pesticides: global use and concerns, in *Silent Spring Revisited*, G.J. Macro, R.M. Hollingworth, and W Durham, Eds. American Chemical Society, Washington, DC, 1987.

Gurthrie, F.E. Pesticides and Humans, in *Introduction to Environment Toxicology.* F.E. Guthrie and J.J. Perry, Eds. Elsevier/North Holland, Amsterdam, 1980.

DOE. *The Non-agricultural uses of pesticides in Great Britain.* Report of the Central Unit on Environmental Pollution. DOE (UK) Pollution Paper No. 3, 1974.

NISOH. *Occupational Disease*. Cinuinnate : US Depth Health and Human Science, 1977.

Sitig, M. *Pesticides Manual and toxic materials control encyclopedia*. Noyes Data Corporation, New Jersey, 1980.

8

Ionizing Radiation

INTRODUCTION

Electromagnetic radiation may be divided into nonionizing and ionizing radiation according to the energy required to eject electrons from molecules. Radiations that are nonionizing have lower energy and are not capable of producing ionization. Examples of these lower energy electromagnetic radiations are ultraviolet, visible light, infrared, microwave, radio wave, electric wave, and sound. Ionizing radiation, (which exhibits the properties of both waves and particles) has enough energy to produce ionization in matter. Those are exhibit particle properties induce alpha particles and beta particles, while those that behave more like waves of energy include X-rays and gamma rays.

First we will consider the factors that influence the toxicity of radiation. One of the major factors related to the exposure is the dose or total amount of radiation received (Table 8.1). The absorbed dose of radiation is the quotient dE/dm where dE is the differential energy deposited into a differential mass, dm. The unit of absorbed dose in the CGS (centimeter-gram-second) system is the rad (radiation-absorbed dose) and 1 rad =100 erg/g; a dose of 1 rad of ionizing radiation has been absorbed when 100 erg of energy have

Table 8.1 : Radiation Quantitites and Units Used in Radiobiology

Unit of Quantity	*Symbol*	*Application*
Becquerel	Bq	SI quantity of radioactivity Bq = 1 disintegration/sec Bq = 2.7×10^{-11} Ci
Curie	Ci	Quantity of radioactivity 1 Ci = 3.7×10^{10} dps 1 Ci = 3.7×10^{10} Bq
Gray	Gy	SI unit of absorbed dose 1 Gy = 100 rad = 1 J/kg
Rad	rad	Unit of absorbed dose 1 rad = 0.01 Gy = 100 erg/g
Rem	rem	Unit of dose equivalent rad × Q × other modifying factor 1 rem = 0.01 Sv
Sievert	Sv	SI unit of dose equivalent rad × Q × other modifying factor, 1 Sv = 100 rem
Linear energy transfer	LET	Energy deposition per unit of path length; usually in eV/μm
Relative biological effectiveness	RBE	Same effect from same dose of reference radiation, used in radobiology
Quality factor	Q	Biological effectiveness of radiations
Electron volt	eV	Unit of energy 1 eV = 1.6×10^{-12} erg 1 eV = 1.6×100^{-19} J

been deposited in each gram of material. Another term commonly used, particularly in the field of radiation protection, is the rem (roentgen equivalent man). This unit was developed to enable radiation protection personnel to set standards of exposure (rem red × quality factor × distribution factor). The quality factor is a unit to equate the relative biological effectiveness (RBE) of one radiation to another, and the distribution factors attempts to compensate for the varying sensitivity of the different parts of the body.

The roentgen, that amount of radiation required to produce 1 electrostatic unit of charge per cubic centimeter of air, is an older radiation exposure term still found in the literature. This is a measure of only the actual ionizations produced by X-ray or gamma ray irradiation in air.

In 1980 the International Commission on Radiational Units and Measurement (ICRU) introduced the "System International" or SI units to express radiation dose. The Gray (Gy), the SI unit for absorbed dose, corresponds to an energy absorption of a j/kg or 100 rad. This concept of energy absorption is useful for determining absorbed doses of X-rays and gamma rays. However, determination of the absorbed dose in tissues exposed to fast neutron radiation involves more elaborate calculations. The absorbed dose of neutron radiation depends on he transfer of energy from neutrons to directly ionizing particles in the tissue and must be described by the kinetic energy released in the material.

For general use, a quantity different from the rad or Gray has been introduced, the dose equivalent. The dose equivalent allows for the relative effectiveness of a particular type of radiation. Gamma rays and X-rays are regarded as the standard and quality factor of 1 is multiplied by the dose to compute the dose equivalent. Therefore, the dose equivalent (Seiverts) for X-rays and gamma rays is equal to the (Grays). However, neutrons are thought to be roughly 10 times more effective in producing tissue damage than X-ray and therefore are assigned a quality factor of 10.

A factor influencing the toxicity of radiation is the dose rate (D=dD/dt, where the differential dose varies with respect to time, or where there is no variability in dose, [dE/dm], D=D/t). When reviewing experiments in radiation toxicology, these variables (i.e., total dose, dose rate, type of radiation, and variability of the model) must be considered.

EXPOSURE TO IONIZING RADIATION

The planet earth is essentially a closed system except for the input of cosmic radiation and debris. Most of us limit our concern of environmental radiation to its damage to humans, perhaps on the beach or in the mountains. However, the quality of environmental radiation could pose a risk to ecological balance. The radiation environment on earth could range from the irreducible natural background levels to extremely high levels following global nuclear warfare. Biological damage may be detected at levels slightly above the former, while the later would make the planet uninhabitable. The source of environmental radiation may be broken into two major components: natural and artificial or technologically induced radiation (Table 8.2).

Table 8.2 : Average Annual Effective Dose Equivalent of Ionizing Radiations

Source	*Dose Equivalent*		*Effective Dose Equivalent*	
	mSv	mrem	mSv	%
Natural				
Radon	24.0	2400	2.0	55.0
Cosmic	0.27	27	0.27	8.0
Terrestrial	0.28	28	0.28	8.0
Internal	0.39	39	0.39	11.0
Total natural	—	—	3.0	82.0
Artificial				
Medical				
X-ray diagnosis	0.39	39	0.39	11
Nuclear medicine	0.14	14	0.14	4.0
Consumer products	0.10	10	0.10	3.0
Occupational	0.009	0.9	<0.01	<0.3
Nuclear fuel cycle	<0.01	<1.0	<0.01	<0.03
Fallout	<0.01	<1.0	<0.01	<0.03
Miscellaneous	<0.01	<1.0	<0.01	<0.03
Total artificial	—	051	0.63	18
Total natural and artificial	—	—	3.6	100

Source : Health Effects of Exposure to Low Levels of Ionizing Radiation (BEIR V). National Academy of Sciences. National Academy Press, Washington, D.C.

Natural Background Radiation

Natural background radiation is the greatest contributor to radiation exposure in the world. In most countries natural background radiation contributes slightly more than half of the absorbed radiation dose. Relative contributions to the total absorbed dose may range from 42% in highly developed countries to about 94% in most developing countries. Exposure to natural sources of irradiation is unavoidable for the most part, and life has evolved under a continuous exposure to ionizing radiation. This background radiation has three components: (a) cosmic radiation (external), (2) terrestrial radiation (external), and (3) naturally occurring radionuclides (internal).

Cosmic Radiation

Cosmic radiation originates predominately from galactic sources and consists mostly of high-energy protons and alpha particles. At the earth surface, cosmic radiation varies with altitude, geomagnetic latitude, and solar modulation.

This effect of attitude becomes increasingly important to passengers and crews of high flying aircraft. It is estimated that cabin attendants and crew members receive about 160 mrem/year above that received at sea level (Eisenbud, 1987). The cosmic rays are attenuated by the earth atmosphere, resulting in a shielding effect which decreases with altitude. Cosmic ray exposure doubles every 1500 m above the earth surface.

Above the earth atmosphere radiation consists of two main components. One is the highly energetic cosmic radiation which is geomagnetically trapped in the earth's magnetic field. The second component is beyond the earth's magnetic field and is due to background cosmic radiation consisting of about 85% protons and 14% alpha particles. Astronauts traveling into outer space must traverse two belts of geomagnetically trapped radiation, the primary cosmic radiation, and radiation from solar flares.

Terrestrial Radiation

Terrestrial radiation levels and rates from natural background sources are functions of geographic location and living habits. In most areas on earth the terrestrial radiation level varies within relatively narrow limits, but in certain regions of Brazil, China, France, Italy, Madagascar, and Nigeria the terrestrial radiation levels substantially exceed the normal range (Eisenbud, 1987).

The conterminous U.S. may be divided into three general radiation region (BEIR III, 1980). The Atlantic and Gulf coastal plains receive an average of 23 mrems/years while the range in the Colorado plateau area may be as high as 140 mrems/year. The average terrestrial level of the remainder of the U.S. is only 46 mrems/years with an estimated national average of 40 mrems/year.

The terrestrial radiation rate varies with the type of soil in the area and the naturally occurring radionuclide content of the soil. Approximately 70 of the 340 nuclides found in nature are radioactive (Eisenbud, 1987). These radionuclides have existed in the earth's crust since its formation and known as "primordial radionuclides". These primordial radionuclides have half-lived comparable to the age of the universe and are the source of terrestrial radiation.

Three distinct chains of primordial radioactive elements are found in the earth's crust and account for much of the terrestrial radiation exposure. These are (1) the uranium series, (2), the thorium series, and (3) the actinium series. Uranium, the origin of the actinium series, is found in various quantities in rocks and soils. The uranium isotopes are alpha emitters and therefore do not contribute to the gamma background radiation. Uranium is soils and in fertilizers can be absorbed by plants and, via the food chain, be subsequently found in animal, tissues. At equilibrium, an adult human male may be expected to have a uranium body burden of 100 to 125 µg. The ^{232}Th decay series may also move through the food, chain, but due to its relative insolubility and low specific gravity it is present in biological material only insignificant amounts (Eisenbud, 1987). Thorium may be found in silty clay and peaty soils and can be absorbed by vegetables such as potatoes, corn, carrots, beans, and squash. However, the principal source of human exposure to thorium is through inhalation of soil particles. Thorium is removed very slowly from bone and its concentration is found to increase with age.

^{226}Ra, an alpha emitter originating in the uranium decay series, is present in varying amounts in all rocks, soils, and water is of special important along with its daughter products. ^{226}Ra, with a half-life of 1622 years, decays to radon, (^{222}Rn), a noble gas radionuclide with a half-life of 3.8 d. Radon also emits alpha particles but adds to the gamma radiation level of the environment through its gamma-emitting daughters.

Radium is very similar to calcium in its chemical properties and is absorbed by plants from the soil in manner similar to calcium. It then passes through the food chain to humans, where 70 to 90% is concentrated in bone. The amount of ^{226}Ra moving through the food chain depends upon its content in the soils and its rate of absorption by plants, which in turn in dependent on the amount of exchangeable calcium in the soil. Brazil nuts, which have a tendency to concentrate barium (another chemical very similar to radium) may have a ^{226}Ra content about 1000 times greater than the average diet.

Internal Radiation

Internal radiation, the third component of natural background radiation, results from naturally occurring radionuclides contained within the body and contributes about 11 to 17% of the average radiation exposure of the population, (BEIR V, 1990). While some of the radionuclides are freely dispersed throughout the body, others are concentrated in specific organs. All of the emitted decay energy from these sources is absorbed locally. The absorption and deposition of naturally occurring radionuclides of bismuth, carbon, hydrogen, lead, polonium, potassium, radium, radon, thorium, and uranium result primarily from the inhalation and ingestion of these materials in air, food, and water (BEIR III, 1980).

In a terrestrial ecosystem radionuclides that occur naturally in soil, or are deposited in the soil, are incorporated metabolically into plants (Eisenbud, 1987). The absorption of radionuclides from soil depends on the chemical form and distribution coefficient of the radionuclide as well as the metabolic requirements of the plant and physicochemical factors in the soil. In addition to root absorption, plants are contaminated by direct foliar deposition. Foliar deposition is potentially a major source of food chain radionuclide contamination since the radionuclide may be absorbed by the plant or transferred directly to animals consuming or coming in direct contact with the foliage.

Atmospheric radionuclides are eventually deposited on surface waters as well as on soil. Therefore, the atmosphere is coupled to soils, surface waters, and subsurface aquifers. Radionuclides are removed from surface soils by processes including surface runoff and leaching into soil water. They are eventually transported into streams or subsurface aquifers. Radionuclides that leach into deep underground aquifers may eventually reach surface waters and become incorporated into the biosphere again.

Rivers, estuaries, and coastal waters are major receptors of effluent radionuclides from industrial plants and cities. These waters are of special importance because of their high biological activity and productivity. Phytoplankton in these relatively shallow waters converts inorganic compounds in aquatic environment into food for higher organisms. Zooplankton, the basic food of several higher trophic levels, uses phytoplankton for nourishment. Certain bottom dwelling fish and animals also consume phytoplankton.

The risk of consumption of radionuclides in marine and fresh water organisms depends, in part, on where the radionuclide is located in the organism. A radionuclide is more of a risk if a concentrated in an organ consumed by higher organisms, such as humans, than if it is deposited in a portion that is not eaten. The radionuclides of cobalt (^{60}Co) and zinc (^{65}Zn) concentrate in edible tissues while those of radium (^{226}Ra) and strontium (^{90}Sr), although concentrated by calms, oysters, scallops, and certain crabs, are stored in the shell which is not ordinarily consumed (Eisenbud, 1987).

Uptake and retention of radionuclides by an organisms are influenced by the portal of entry, the chemistry and solubility of the compound, particles size, and metabolism. Uptake normally occur via three principal routes of entry: inhalation, ingestion, and skin absorption. Of the three, inhalation poses the greatest risk. Particles greater than 10 mm in diameter do not penetrate into the lung. These particles are removed by the nasal hairs or cilia found in the upper respiratory tract. Therefore, particles size will determine the compartment into which the particle is deposited.

Ingestion of contaminated food contributed to exposure. Gastrointestinal (GI) exposure depends upon transit through the gut, while absorption depends upon the solubility of the radionuclide. Contamination of skin with radionuclides is of less consequence since the skin forms a formidable barrier. However, contamination of an open would may result, not only in continuous radiation of the surrounding tissue but in the introduction of the radionuclide into the rest of the body.

Regardless of the portal of the entry, the radionuclide passes throughout the body of the animal and is deposited into the milk, flesh, and eggs. When the radioactive material enters the body it becomes an internal emitter. It will continue to radiate the body until it is excreted by some physiologic process, mainly in the urine and feces, or until its radioactivity decays. The time it takes an organism to eliminate half of the radionuclide is known as the *biological half-life* and the time required for a radionuclide do decay half of its activity is the *physical half-life* (Table 8.3).

Radon

Although not mentioned specifically in the above discussion of internal radiation from radionuclides, radon (^{222}Rn), the short-lived decay product of ^{226}Ra, may be the most important internal radionuclide, accounting for about 60% of the effective dose equivalent from internal emitters. As seen in Table 8.2, radon and its decay products contribute 55% of the total average annual effective dose equivalent of 3.6 mSv.

^{222}Ra is a naturally occurring, colorless, odorless, tasteless, radioactive gas formed from the decay of uranium and radium. Since uranium has been present since the earth was formed and has a half-life of 4.5×10^9 years, radon has also been present for some time will likely exist for many years at the same level as it is now.

By the early part of the 20th century radon was being linked with lung cancer while being used therapeutically. In the 1960s

Table 8.3 : Half-Lives of Some Biologically Significant Radionuclides

	Half-Life		
Nuclide	Physical	Biological[a]	Effective
^{14}C	5.73×10^4y	40 d	40 d
^{137}Cs	30 y	70 d	70 d
^{131}I	8 d	12 d	8 d
^{55}Fe	657 d	2000 d	494 d
^{32}P	14 d	260 d	14 d
^{239}Pu	2.4×10^5y	180 y	180 y
^{24}Na	15 h	11 d	14 h
^{90}Sr	29 y	36 y	16 y
^{3}H	12 y	12 d	12 d
^{235}U	7.1×10^8y	20 d	15 d
^{65}Zn	245 d	400 d	152 d

[a] Whole body

Source : Eisenbud, 1987.

attention was focused on the emission of radon from uranium mill tailings in parts of Colorado and Utah. An enterprising contractor was using these mill tailings for aggregate in the constructions of houses. However, since uranium is not mined in many places, concern regarding its use did not remain in the national spotlight for very long.

Production

Production of radon is primarily from the widespread distribution of uranium and its decay products in the soil. Every square mile of surface soil to a depth of 6 in. contains about 1 g of radium, a decay product of uranium. Decay of radium releases radon in small amounts to the atmosphere. The radium content of the soil reflected the radium content of the rocks from which it was formed, and radon released to the atmosphere is increased in areas with granite formations and uranium and thorium ore deposits.

Mineral waters were thought to have some curative powers in Roman times, and it is known that many mineral springs contain relatively high concentrations of radium and radon (Eisenbud, 1987).

Potable water supplies may also contain radon due to the deposition of radium isotopes. Groundwater supplies may contribute significantly to indoor radon concentrations. Private groundwater supplies constitute a somewhat greater source of radon than do public supplies.

Commercially produced radon has been used by radiopharmaceutical companies and hospitals to produce radioactive seeds or needles which are then implanted into tumors. For the most part, the use of radon in this procedure has been replaced by radionuclides made in accelerator and nuclear reactors.

Use

Each year thousands of people throughout the world use radon in some form for therapeutic purposes. Ailments treated in various ways include acne, allergies, arthritis, asthma, diabetes, gout, hypertension, and ulcers. The National Health System in the U.S.S.R. prescribed radon bath treatments daily. Medical uses of radon to treat malignancies began in the U.S. the early part of the 20th century and was used in the treatment of dermatological disorder as late as 1950.

The emanation of ^{222}Rn from soil and its concentration in groundwater have been used more recently as good predictors of earthquakes, for the study of atmospheric transport, and for exploration of petroleum and uranium deposits.

Environmental Entry

^{226}Ra in soil is the largest single source of radon in the atmosphere. Radon is continually being formed in the soil and released into the air at about 2×10^9 Ci ^{222}Rn/year. The rate varies from one location to another depending on many factors including soil composition, ambient temperature, and barometric pressure. A small amount of ^{222}Rn is absorbed by plants and then released into the air.

When ^{222}Rn enters the atmosphere from the ground it is dispersed into the air and its concentration in open areas is low. However, when ^{222}Rn enters a building, either from the building materials, the soil beneath the floor, or around the walls, the concentrations builds up because of the restrictions in air movement within the building and lack of air supply from the outside. The Rn and its decay products with short half-lives may adsorb to dust particles in the air and be inhaled.

Groundwater that is in contact with rock or soil containing radium will pick up the ^{222}Rn and Release it to the atmosphere when the water comes to the surface. With a contribution of 5×10^8 Ci ^{222}Rn/year groundwater is considered the second largest source of environmental radon. Tailing from uranium mines, coal residues, natural gas, and some building materials release small amounts of radon to the environment.

Environmental Levels

Although the units of radiation and absorbed dose were defined earlier, the measurement of radioactivity was not discussed. The unit of radioactivity of many years has been the ***curie (Ci), which is defined as the quantity of radioactivity producing 3.7 × 10^10^ nuclear disintegrations per second***. The System Internationale introduced in 1980, defined new terms for radioactivity and radiation dose. The newer ***SI unit*** being substituted for the curie is the ***becquerel (Bq), with I Bq being equal to one disintegration per second.***

Not only are radioactivity levels of ^{222}Rn written as pCi/L or Bq/m^3, but as *working level* (WL) or *working level month* (WLM). The WL for radon exposure is defined as any combination of short-lived radon daughter radionuclides in 1 L of air that will result in the emission of 1.3×10^5 MeV of potential alpha energy (BEIR IV, 1988). This corresponds to a ^{222}Rn concentration of 3.7 kBq/m^3. Exposure to one WL for a month of 170 working hours will result in one WLM. However, a 30-d month is equal to 720 h and exposure to one WL for 720 h gives a cumulative exposure of 4.235 WLM. Therefore, remaining at home for 12 h/d for a month at an exposure level of 1 WL would result in an exposure of greater than 2 WLM/month.

Typically, the ^{222}Rn concentration in air over soil is 4Bq/m^3 (100pCi/m^3), but this value may vary by a factor of two or more with seasonal and diurnal variations (BEIR IV, 1988). Seasonally, levels are highest in early autumn and lowest in early spring, while the diurnal variations show an early morning peak and a decrease in the afternoon.

Distribution studies of household levels showed that 7% of the U.S. single-family houses have concentrations greater than 150 Bq/m^3 and that possibly a million houses in the U.S. have annual average ^{222}Rn concentrations of 300 Bq/m^3 or more. Average winter concentrations found in houses in the Spokane Rier Valley of Washington and Idaho and in eastern Pennsylvania were on the order of 500Bq/m^3. Some houses in eastern Pennsylvania had levels as high as 100,000 Bq/m^3, and in one neighborhood of Clinton, New Jersey, winter concentration exceeded 7500 Bq/m^3.

^{222}Ra levels in surface water are essentially zero. However, levels in aquifers of igneous and metamorphic rocks are high. Aquifers associated with granite consistently show the highest levels of ^{222}Rn, averaging 100 Bq/L.

Radon in soil is usually referred to as soil-gas and the levels in soils are affected primarily by the radium content and its distribution in the soil, soil porosity, moisture, and density. Measurements of

^{222}Rn soil-gas have been minimal but those reported range from 7000 Bq/m^3 in Spokane, Washington to 1,000,000 Bq/.m^3 in Reading Prong. New Jersey. Plants apparently absorb ^{222}Rn from soil but little information concerning this process is available.

Environmental Fate

Regardless of where it is found, the ultimate fate of ^{222}Rn is degradation by radioactive decay. ^{222}Ra, with a half-life of 3.82 d, degrades by alpha-emission to ^{218}Po. The radon daughters are electricity charged and attach to inert dust in the atmosphere, resulting in the formation of radioactive dust. Because most of the daughters are short-lived, with ^{210}Pb having the longest half-life of 22 years, equilibrium is reached in the air in the about 22 h (Eisenbud, 1987). The concentration of ^{222}Rn daughters in the atmosphere can be decreased by the passage of air containing dust over the ocean or by a thunderstorm. In any event, the mean life of the atmospheric dust with attached radon daughter is thought to be 15 d. However, radon and daughters are transformed only though radioactive decay, emitting alpha particles in the process.

Technologically Induced Radiation

Health Sciences

The use of man-made radiation in the health science is normally divided into three areas: (1) diagnostic X-ray examination, (2) nuclear medicine, and (3) therapeutic radiation. The use of X-rays in diagnostic examinations, including dental, represents the single largest man-made source of radiation exposure in the U.S. (BEIR III, 1980). In the more highly developed countries such as the U.S. exposure from medical sources may even exceed natural background exposure. Dental X-ray are the most common of the diagnostic examinations with outpatient examinations contributing 30% of the exposure.

The use of radiopharmaceuticals in nuclear medicine has almost doubled over a 10-year period. It is estimated that up to 12 million doses of radiopharmaceuticals are dispensed each year in the U.S. for diagnostic purposes (BEIR III, 1980). However, the pet capita effective dose equivalent from these procedure in the U.S. is only about 140 μSv.

Radiation therapy has been used almost exclusively for the treatment of malignant neoplasms. The hgh absorbed dose, 50 to 70 Gy, required in most malignant conditions leads to *nonstochastic* or direct effects such as cell death. Therefore, normal tissue surrounding the neoplasm may also be exposed and incur long range risk. The risk, however, may be eclipsed by the immediate benefits associated with increased life expectancy resulting from the destruction of the neoplasm.

Nuclear Power Production

When radiation exposure from nuclear power production is mentioned, the immediate thought is of nuclear power reactors and the environmental dispersion of radionuclides, particularly ^{85}Kr, ^{3}H, ^{14}C, and ^{129}I. However, exposure from nuclear power production should also include mining, uranium fuel fabrication, and waste storage and disposal.

Although uranium mining increases the amount of uranium and its decay products, including radon and its daughters, the environmental risks from the radioactive emissions from uranium mines is insignificant. (Eisenbud, 1987). However, mill tailing may represent a significant source of environmental radiation due to the emanation of ^{222}Rn, dispersion of the tailings by wind and water, and by the use of mill tailings in building construction.

About 1000 nuclear reactors have been constructed and are operational throughout the world. Some of the reactors were built for research or the production of radioisotopes and plutonium. Approximately 200 naval vessels throughout the world are powered by nuclear reactors. Yet, the environmental release from nuclear operations in the U.S. results in a dose rate for the average person of <1 mrem/year (BEIR III, 1980).

Nuclear Weapons

The first atomic weapon was detonated on a July morning in 1945 on a New Mexico desert north in Alamagordo. Since that day, hundreds of test explosions have been conducted by the U.S., the U.S.S.R., the U.K., India, France, and the People's Republic of China. In 1963 the U.S., the U.K., and the U.S.S.R. signed an agreement to end atmospheric testing. However, France, China, and India did not sign this agreement and continued to conduct atmospheric testing (Eisenbud, 1987). Between 1945 and 1984 the total estimated yield of all atmospheric nuclear explosions was about 546 Mton (Mettler and Moseley, 1985). A 1 Mton explosion equals the explosive force of 1 x 10 ton of TNT.

The detonation of a nuclear weapon generates enough heat to vaporize the weapon instantly and stop the nuclear reaction. The nuclei created by the reaction are now highly energetic and emit high-energy radiation such as X-rays and gamma rays as they return to a lower energy state. The released radiation heats the surrounding air, forming a shock wave and a fireball. For a nuclear weapon with a yield of 1 Mton, the fireball will rise at a rate of almost 400 ft/sec and reach an attitude of about 60,000 ft. Convective forces initiated by the fireball will result in enormous amounts of air, soil, and debris being sucked upward, creating crater as much as 400 ft deep and 1200 ft in diameter. As the fireball cools, the newly created, highly unstable nuclei condense onto particles of soil and debris, returning to earth as radioactive fallout.

In a nuclear explosion, over 400 radioactive isotopes are released into the biosphere. Among these, about 40 radionuclides are considered potentially hazardous. Of particular interest are those isotopes whose organ specificity and long half-lives present a danger of irreversible damage or induction of malignant alterations. Both early and delayed fallout result in the deposition of radioactive material in the environment.

The radioactive fallout from nuclear explosions may be divided into three portions depending on the yield and height of the burst. The larger, intensively radioactive particles fall close to the site within hours. Slightly smaller particles behave somewhat like aerosols and are dispersed into the troposphere where they will stay for several months. The fallout from this portion remains in bands around the earth at the latitude of the detonation. The third portion penetrates the stratosphere and its particles are deposited worldwide over a period of months to years (Eisenbud, 1987). Most of the radioactive fallout is down wind from the explosion and upto 70% is comprised of the largest particles, returning to earth close to the detonation site within hours. The intensity of the radioactivity varies inversely with distance from the site of explosion. With a steady wind, the pattern of accumulated dose of radioactivity assumes the shape of nested cigar- shaped contours each contour denoting a particular dose.

A 1 Mton thermonuclear weapon, detonating at ground level with a steady wind of approximately 15 mi/h, would produce a radioactive fallout dose rate of 400 rem in 24 h in an area of approximately 400 sq mi. At a dose rate of 2 rem/year, more than 20 times the maximum recommended by the EPA, an area of 1200 sq mi would remain unfit for use for a year and more than 20,000 sq mi would be uninhabitable for a month.

Accidents

Although the environmental release of radionuclides from nuclear reactor operations is about 1 mrem/year per person, malfunctions can occur and accidents can happen (BEIR III, 1980). On March 28, 1979 the worst accident in the history of U.S. commercial nuclear power generation occurred on Three Mile Island in Pennsylvania. Although radiation exposure to the plant workers and the public was insignificant, the nuclear power industry was set back almost a decade. Orders for the construction of new nuclear plants were canceled (Eisenbud, 1987).

This is not to say that all nuclear power plant accidents will not result in the discharge of radionuclides. Not all of the nearly 800 nuclides produced in the reactor are radioactive and of these only 54 considered a significant risk (Eisenbud, 1987). With core damage, the severity of the accident and therefore the risk, depends mainly on the release of radioactive isotopes, mainly ^{131}I and ^{137}Cs, into the environment.

Since 1952 there have been 14 reactors accidents that involved core damage. One, the Windscale, U.K. Atomic Energy works accident, was the first time radioactive material was released from a reactor accident. In October 1957, a plutonium production reactor located on the coast of Cumbria in northwest England released about 20,000 Ci ^{131}I (Eisenbud, 1987). The core in one of the two air-cooled, graphite-moderated, natural uranium reactors was partially consumed by fire, releasing the fission products onto the seashore and foothills southwest of the Cumbrian Mountains. As an aftermath of this accident four fatal leukemia cases in people that were under 20 years of age between 1950 and 1980 were recorded in the nearby village of Seascale. Based on statistical epidemiological calculations, only 0.5 cases would have been expected.

At 1:23 a.m. on Saturday, April 26, 1986, an explosion

occurred at water-cooled, graphite-moderated reactor of Unit No. 4 at Chernobyl power station along the Pripyat River about 90 km north of Kiev, Ukraine. This accident resulted in the release of an estimated 81 MCi of radioactivity into the atmohphere by May 6, 1986 and was the most costly industrial accident in history (Eisenbud, 1987).

The radioactive cloud, because of wind direction, moved northwest from Chernobyl, covering much of Scandinavia by Monday, April 28 (Clarke, 1987). On the Baltic coast of Sweden, the radiation levels were 14 times the normal background level. By Wednesday, April 30, the wind direction at Chernobyl had shifted and part of the radioactive cloud began moving to the southeast. A high pressure also developed over eastern Europe, splitting the cloud into two parts. The western portion of the cloud covered all or part of Scandinavia, Poland, Germany, Austria, Czechoslovakia, Hungary, Switzerland, Italy, and Yugoslavia. By Friday, May 2, the eastern portion of the cloud had covered all or part of Romania, Bulgaria, Yugoslavia, Greece, and Turkey, while the western portion had moved over part of France and the U.K. The Netherlands and Belgium were completely covered by this time.

One week after the accident two masses of contaminated clouds still existed, one over northwestern Europe and the other over southeastern Europe. By Monday, May 5, the largest contaminated cloud lay over southern Germany, most of Italy, Greece and eastern Europe. The northwestern cloud was then dispersing over the Atlantic.

The accident at Chernobyl released the quantities of various radionuclides in the atmohphere, resulting in widespread fallout throughout the northern hemisphere. This, is turn, led to a large increase in measured values of range of radionuclides in various environmental domains. In all cases the most prominent radionuclides were ^{131}I, ^{134}Cs, and ^{137}Cs. The highest levels of ^{131}I and ^{137}Cs fallout were detected in southern Germany, Austria, northern Italy, and Greece (Clarke, 1987). The radionuclide deposition was dependent upon meteorological conditions during passage of contaminated air. In the U.K. high deposition occurred in areas where it rained during passage of the contaminated air mass. By a strange twist of fate, these areas included north Wales, southwest Scotland, and the Lake District of England. This latter area of England was the site of the 1957 Windscale accident.

In comparing the size of the releases and radiological consequences of the 1957 Windscale accident, the Three Mile Island (TMI) accident in 1979, and the Chernobyl accident in 1986, a striking difference in the impact among the three accidents was noted (Clarke, 1987). While the TMI Release was higher than at Windscale, the effective dose was much lower. This was due to the radionuclides released, ^{131}I at Windscale and ^{133}Xe at TMI, and that the fission products at TMI were retained within the vessel. The release from the Chernobyl accident was 10 times higher than from TMI but the doses were hundreds of times higher than from TMI. Again, this was partially due to the radionuclides released at Chernobyl (^{131}I and ^{137}Cs). The total impact of the Chernobyl accident was greater in the U.K. than was the Windscale accident, while the total impact of the TMI accident was 100 times less in the U.S. than the Windscale accident was in the U.K.

Four years after the Chernobyl accident Soviet diplomats acknowledged that the medical, environmental, and political consequences are much greater than previously discussed. Four million people in Ukraine, Byelorussia, and western Russia are said to be living on contaminated ground, and the death rate of children with leukemia in Minsk is 50 times greater than a year before the accident.

Waste Management

The disposal of radioactive waste is a dilemma facing a technologically advanced society. One of the basic demands of such a growing society is availability of convenient and inexpensive sources of energy. As the demand for energy increases, the reliance on nuclear power will increase, as will the production of radioactive waste. The long-term storage and disposal of high-level nuclear waste is a problem that has not been adequately resolved.

Radioactive waste is classified according to its physical and chemical properties as well as its sources (Eisenbud, 1987). The three general categories of radioactive waste are

1. Low-level waste
    a. Residues from laboratory research
    b. Uranium mill tailings
    c. Waste generated in the cleanup of uranium, radium, and thorium processing plants

2. High-level waste
   a. Spend fuel from civilian nuclear power reactors
   b. Liquid and solid residues from the reprocessing of civilian spent fuel.
   c. Liquid and solid waste and reprocessing of fuel used for military purposes.
3. Transuranic waste — mainly alpha-emitting residues from military manufacturing.

Low-level radioactive waste management is facing a crisis, not because of technical problems and environmental risks involved, but because of the socio-political ramifications of the widespread public concern of the perceived risk of shallow and land burial (Eisenbud, 1987). Although damage to public health has not been shown to have resulted from operations at six commercial shallow land burial sites and five major government shallow land burial facilities, three of the commercial sites have been closed. Of the three remaining commercial low-level sites in the U.S., the Chem-Nuclear Systems site at Barnwell, South Carolina is the only commercial site in the east and it receives more than half of the low-level waste generated in the U.S. Since the national inventory of low-level radioactive waste is growing at about 10^5 m^3/yr (30% from medical institutions), some other method must be employed to dispose of low-level radioactive waste.

Of greater concern is the storage of large quantities of highly toxic liquid and solid wastes which must be isolated from the environment for thousands of years. The selection of sites to store high-level radioactive waste requires more than technical decisions. Each site that is technically acceptable must also have public support. There is general agreement by the public that there is a need to store radioactive waste. However, when a site is found which meets the technical qualifications, the local public wants the waste stored somewhere else. Eisenbud (1987) calls this a "not in my backyard" (NIMBY) syndrome.

Much of the high-level nuclear waste is stored at temporary sites in concretencased steel tanks a few meters below the surface of the ground. These tanks have a capacity ranging from 15,000 gal to 1 × 10^6 gal and an expected lifetime of 15 to 40 years (Esenbud, 1987). Considering the finite lifetime of the steel tanks, it becomes clear that these wastes must be transferred to other containers or more secure sites in the future.

Scientists appointed by the International Council of Scientific Unions believe that interim storage for 50 to 100 years will greatly reduce the problem of thermal loading when the waste is finally transferred to the final disposal site. They believe that current and future technology will allow the safe disposal of the nuclear waste. Several options for the permanent management of radioactive waste have been considered (Eisenbud, 1987). The major methods being considered for disposal or long-time storage include (1) as solids in salt mines, deep underground caverns, man-made vaults, deep ocean seabeds, deep ocean subseabeds, deep mined cavities, or insertion into Greenland glaciers; (2) as liquids in deep wells, deep underground caverns or in tanks; and (3) as a grout mixture injected into deep rock fissures. The disposal method now in favour is deep underground mined cavities.

As a result of the Nuclear Waste Policy Act signed into law on January 7, 1983, the U.S. Department of Energy (DOE) has proposed five preliminary sites for chacaterization. These five sites are located either in basalt, bedded salt, domal salt, or tuff geological media found in Richton Dome, Mississippi, Yucca Mountains, Nevada, Deaf Smith County, Texas, Davis Canyon, Utah, or the DOE Hanford Site in Washington. January 31, 1998 has been established as the date when the first repository operation will begin.

BIOLOGICAL EFFECTS OF IONIZING RADIATION

Radiation toxicology is a specialized branch of toxicology involving the study of the adverse effects of radiation on living organisms. It is a multi-disciplinary science, borrowing freely from several of the basic sciences. The cytopathologic effects of radiation exposure are quite similar to those induced in other types of cellular injury. Radiation-induced cell changes may result in death of the organism, death of the cells, or cancers with no features distinguishing them from those induced by other types of cell injury.

Radiation exposure occurs from many of the sources mentioned previously. *Directly ionizing* radiation carries an electric change that interacts directly with atoms in the tissue or medium by electrostatic attraction or repulsion. *Indirectly ionizing* radiation is not electrically charged, but results in production of charged particles by which its energy is absorbed. A characteristic of charged particles produced directly or indirectly is *linear energy transfer* (LET), the energy loss per unit of distance traveled, expressed in kiloelectron volts (keV) per micrometer (μm). The LET, depends on the velocity and charge of the particles produced and varies from about 0.2 to > 1000 keV/μm.

Many particles spend virtually all their energy at LETs of less than a few keV/μm. The most significant of these particles are the principal components of primary cosmic radiation and high-energy electrons, such as those emitted by beta radiation. These high energy electrons as well as the indirectly ionizing radiation such as X-rays and gamma rays that produce them, are referred to as low-LET radiation. Low energy electrons, produced by both direct and indirect ionizing radiation, are intermediate in LET.

High-LET radiation also contributes to the environmental radiation load. Alpha radiation emitted by internally deposited radionuclides is probably the most important directly ionizing high-LET radiation. Neutron radiation is the principal indirectly ionizing high-LET radiation.

Irradiation induces formation of reactive chemical products when it enters a biological system. For example, superoxide radicals can be generated in the reaction of hydrated electrons or hydrogen atoms with dissolved oxygen following gamma radiation of aqueous solutions. This radiolytically-formed free radical is involved in oxidative chain reactions (Fridovich, 1976) with the possibility of interconversion and postirradiation generation of other forms of activated oxygen, leading indirectly to further irradiation-induced cellular damage (Greenstock, 1984).

Low-level ionizing radiation of living cells results in the formation of free radicals and hydroxyl radicals. Much of the cellular DNA damage inflicted by ionizing radiation is due to the formation of free radicals and hydroxyl radicals. Like-wise, the formation of hydroxyl radicals by ionizing radiation can initiate lipid peroxidation, leading to cell membrane damage and the formation of cytotoxic aldehydes.

Ecotoxicology

Plants and lower animals are much more resistant to the effects of radiation than humans (Eisenbud, 1987). Nevertheless, when a mixed forest on Long Island was subjected to chronic gamma irradiation during the growing season for 12 years. a phenomenon known as retrogression occurred. The results of the irradiation of an oak-pine forest by a radiation source containing 9500 Ci ^{137}Cs have been documented. Within 6 months a vegetation gradient developed. What was originally on oak-pine forest become an oak forest, with the elimination of the pine. The oak forest then became a shrub zone, followed by a sedge zone. Near the radiation source, the sedge gave way to a central devastated area, where only mosses and lichens survived. The changes were similar to those seen along gradients of increasing severity of climatic changes, such as increasing altitude on a mountain. The chronic exposure to gamma radiation reversed the ecological succession.

After 12 years of chronic irradiation, only lichens and green algae survived up to a distance of 20 m from the source, an area receiving as much as 3000 roentgen (R)/d. Only sedges and grass lived from 20 to 75 m from the source, where the dose was 20 to 160 R/d. From 75 to 85 m, with a dose of 10 to 20 R/d, blueberry bushes died and no oaks survived. From 85 to 115 m from the source, where 5 to 10 R/d were received, oaks survived but pines did not. Scraggly, stunted pines were seen at a distance of 125 m, where 2 R/d of gamma radiation was received. Even as far away as 150 m from the source, the pine needles were shorter than normal and the diameters of the trees were reduced.

The only animals left on the Long Island experimental plot were insects. Leaf-eating insects attacked the oak trees and shrubs. Bark insects and wood borers developed larger populations in response to an abundance of food and an absence of predators. All other animals died or left the denuded areas.

These observations agree with other studies (Odum, 1971) on the effects of gamma radiation on whole communities and ecosystems in a tropical rain forest of Puerto Rico and in the Nevada desert. The effects of mixed gamma/neutron radiation have been studied on fields and forests in Georgia and at the Oak Ridge National Laboratory in Tennessee. Short-term effects of gamma radiation have been studied at the Savannah River Ecology Laboratory in South Carolina. The Oak Ridge National Laboratory has also conducted a low-level chronic radiation study of a lake bed community.

In all of these studies, the differential sensitivity of a species in an ecosystem is of considerable interest (Odum, 1971). If an ecosystem receives a higher level of radiation than was confronted in its evolution, the elimination of sensitive strains or species may result.

Effects on Plants

In gamma source experiments, such as the ones mentioned above, weeds and grasses proved relatively tolerant of radiation and field crops were only slightly more sensitive. Exposure to sustained irradiation from a large gamma source is not an exact simulation of chronic exposure to short-lived radioisotopes, but it does provide an indication of the effects of an acute, high-level radioisotope exposure.

Radionuclides such as ^{210}Ra, ^{226}Ra, and ^{222}Rn that occur or are deposited in the soil are translocated by plants. In addition to root uptake radioisotopes are absorbed following direct deposition on foliar surfaces. Although the mechanism of radionuclide adsorption by plants is not well understood, it is known that individual radionuclides are translocated from the root or the leaves to the remainder of the plant. Mean ^{232}Th concentrations of 0.018 ± 0.022 pCi/kg have been found in the edible portions of 25 vegetables, including beans, carrots, corn, potatoes, and squash. Flora near the summit of the Morro do Ferro, a hill in the state of Minas Gerais, Brazil have absorbed so much ^{228}Ra that they can be autoradiographed easily (Eisenbud, 1987).

Individual plant species vary widely in their susceptibility to the damaging effects of ionizing radiation. Exposure of the mixed vegetation of a Long Island forest to gamma rays demonstrated the vulnerability of conifers compared to the radiotolerance of grasses. Exposure in early stages of development or during the growing season, when cells are dividing, increases the radiosensitivity of the plant. During cell division, cell death or damage is induced at much lower irradiation levels than those required during the interphase stage of the cell cycle. Investigations of plant damage from gamma ray sources demonstrated that radiation causes a reduction in the number of cells per meristem (growing point). This reduction in cells varies directly with the number of cells displaying chromosome damage.

In higher plants, sensitivity to ionizing radiation is directly proportional to the chromosome volume of the cell nucleus (Odum, 1971). However, certain community attributes such as biomass and diversity also become determinants of species vulnerability. Indeed, the "unshielded" biomass above ground is a major determinant in that plants sprouting from seeds or from shielded under-ground parts have an increased chance of recovery.

Effects on Animals

In higher animals, unlike in plants, there is no simple, direct relationship between nuclear chromosome volume and sensitivity to ionizing radiation. Rather, the effects of irradiation on specific organ systems are more critical (Odum, 1971).

Developmental Stages

As in plants, proliferating cells are much more sensitive to irradiation than differentiated, non-dividing cells. An organism is more radiosensitive during its early stages of development, when most cells are dividing, than at any other stage of its life. The LD_{50} for fish embryos is 16 to 18 times smaller than required in later life. For insects, the LD_{50} in adults in about 1000 times larger than during the developmental stages.

Instead of lethality, morphological abnormalities are associated with irradiation during the middle stages of development. Irradiation-induced anomalies occurring during this time may result in death of the organism or abnormal development of one or more organ systems. Exposure during this period may result in gross malformations, growth retardation at term or as an adult, and structural neuropathology. In the human, most major organogenesis occurs during the first trimester of pregnancy, with embryonic death and congenital abnormalities resulting from irraidation exposure during this period.

During late organogenesis and in the perinatal period, just before and just after birth, radiation damage tends to be functional rather than structural. Perinatal irradiation with X-rays and gamma rays (140 to 180 cGy) induces changes in tissue enzyme activity.

Reproduction

During a period of about 20 years in the early part of this

century, radiation was used in an attempt to increase fertility. Exposure normally was to 1.5 to 2.25 Gy over a period of 3 weeks. These levels apparently had little effect on fertility or on any later conceived children.

In recent years, the safety of radiation levels is more often questioned since the threshold for the effects of ionizing radiation on male reproduction is difficult to predict. Although fully developed sperm cells and primary spermatocytes are relatively radioresistant, the proliferating spermatogonial cells of the testis are highly sensitive to ionizing radiation.

Occupational radiation exposure of the testis produces a significant decrease in serum gonadotropins and significant changes in semen production and morphology. However, the effects of ionizing radiation on spermatogenesis are normally reversible with the recovery of fertility predictable. Although an acute irradiation dose of 6 Gy to the testis is likely to produce permanent sterility, conception has occurred for males after years of either aspermic or hypospermic conditions following absorbed doses between 2.3 and 3.7 Gy (Mettler and Moseley, 1985; Robertson, 1989).

Genetic Effects

Mutations are structural changes occurring in genes as a result of exposure to a number of environmental agents including chemicals, heat, and ionizing radiation (Eisenbud, 1987). Partially due to the efforts of the film industry, its post-World War II movies portray inaccurately the genetic effects of irradiation. In fact, the genetic effects of irradiation dominated the philosophy of radiation risk and radiation protection. Only during the past two decades has the scientific community increased its appreciation of the somatic effects of ionizing radiation. Data evaluation during this time also suggested that the previous estimates of genetic risk were too high.

Damage to DNA along a low-LET radiation tract is the same kind that occurs spontaneously. It will likely consist of single strand breaks in the double helix and may be repaired by cellular enzymes. The ultimate effects depend more on the effectiveness of the repair than upon the initial break. Chromosomal breaks from low-LET radiations have a small probability of resulting in a translocation during repair of a lesion. In contrast, high-LET radiation lesions have a high probability of interaction and result in a greater number of unrepaired or misrepaired lesions. These lesions may be amplified many times during transcription and translation and are the major contributors to the genetic damage resulting from irradiation.

Current studies indicate that there may have been one possible mutation in the 78,000 children of people who survived the atomic bombings at Hiroshima and Nagasaki. When the mutation rate of the group of exposed parents is estimated, this single mutation has a high probability of being unrelated to radiation exposure.

Available information suggests that the radiation dose required to double the human mutation rate varies between 0.8 and 2.4 Gy (UNSCEAR, 1988).

Radiation Carcinogenesis

Radiation carcinogenesis is considered by most radiobiologists as the most important effect of exposure to levels of ionizing radiation below 1 Gy. Ionizing radiation increases the incidence of virtually every type of neoplasm, whether benign or malignant, and the carcinogenic effects of radiation have been observed in practically all species.

The development of cancer appears to be a multistage process involving an *initiator* and at least one *promotor*. Promotors may elicit the production of activated forms of oxygen, including superoxide radicals and peroxides, and hydroxyl radicals, which either directly or indirectly affect DNA. Certainly, radiation causes strand breaks and modification of DNA bases. It is also well known that radiation of living cells elicits the formation of free radicals, and hydroxyl radicals, and it is well established that hydroxyl radicals are responsible for the radiation-induced DNA damage.

At times it becomes difficult to characterize ionizing radiation as either initiator or promotor. Fòr example, cigarette-smoking uranium miners exposed to radon have radiation-induced lung cancer at five times the rate of non-smoking miners. It is difficult to assign the role of initiator or promotor to either radon or cigarette-smoke since both have been implicated in the etiology of lung cancer.

Acute Radiation Syndrome

There have been fewer than 25 documented fatalities worldwide between 1946 and 1985 that can be attributed to radiation accidents. Although exposure of the whole body to lethal amounts of ionizing radiation is very rare, any discussion of the biological effects of ionizing radiation would be incomplete without mentioning acute radiation sickness and acute radiation syndrome (ARS).

Acute radiation sickness is manifest in characteristic clinical sequelae known as ARS, a combination of syndromes determined primarily by the total radiation dose received, the rate of exposure, and the distribution of the radiation in the body. Signs and symptoms of ARS result from injury to bone marrow, gastrointestinal system, cardiovascular system, CNS, gonads, and skin. The variation in radiation sensitivity of these tissues causes the signs and symptoms of ARS to occur in three successive phases: an initial prodromal phase. a subsequent latent period and the manifest illness phase The length of each phase may vary directly with the radiation dose, and the time between each phase may vary indirectly with the dose, so at an extremely high dose of radiation, the phases will blend with the latent period disappearing completely (Table 8.4).

The initial prodromal phase is characterized by a combination of gastrointestinal and neuromuscular symptoms such as anorexia, nausea, vomiting, diarrhea, apathy, tachycardia, fever, headaches, insomnia, dizziness, and vertigo. The pathogenesis of the prodromal phase is not known, but several causal factors have been suggested, including direct radiation effects on the central and autonomic nervous systems, disturbance of endocrine balance, and the production and release of various chemical mediators.

The manifest illness phase of the ARS is classically divided into three major syndromes tradiationally known as the hemopoietic syndrome, gastrointestinal (GI) syndrome, and central nervous system (CNS) syndrome. However the current view replaces the CNS syndrome with the neurovascular syndrome.

The hemopoietic syndrome occurs following exposure to 200 to 700 cGy. Radiation doses of 100 cGY or more can significantly damage the blood-forming capability of the body. Approximately 50% of individuals exposed to 300 cGY will die within 2 months. The signs and symptoms result from radiation damage to the bone marrow, lymphatic organs, and immune system. The pathophysiological effects of damage to bone marrow include increased susceptibility to infection, bleeding, anemia, and lowered immunity. Death usually results from hemoorrhage and infection.

At radiation doses of 7 to 50 Gy, injury to the GI tract inhibits the renewal of the cell lining. The intestinal epithelial stem cell is the target of radiation damage; and the resulting decrease in mitotic activity leads to denudation of the intestinal mucosa, fluid and electrolyte imbalance, and bacteremia. The symptoms of the GI syndrome include lethargy, diarrhea, dehydration, and sepsis. At doses of 3 to 8 Gy, temproary injury to the tight junctions between epithelial cells of the mucosal lining permits the escape of bacterial endotoxins into the bloodstream. As the dose increases, the epithelial lining is more extensively depleted. With doses of 10 to 15 Gy, denudation of the mucosa exacerbates the loss of fluid and electro-

Table 8.4 : Radiation Dose Ranges and Associated Pathophysiological Events

	Pathophysiological Events		
Dose Range (cGy)	Prodromal Effects	Manifest-Illness Effects	Survival
75–150	Mild	Slight decrease in blood cell count	Virtually certain
150–300	Mild to moderate	Beginning symptoms of bone marrow damage	Probable (>90%)
300–530	Moderate	Moderate to severe bone marrow damage	Possible — bottom third of range : $LD_{5/60}$ middle third: $LD_{10/60}$ top third : $LD_{50/60}$
530–830	Severe	Severe bone marrow damage	Death within 3.5 to 6 weeks bottom half: LD90/60 top half: LD99/60
830–1100	Severe	Bone marrow pancytopenia and moderate intestinal damage	Death within 2 to 3 weeks
1100–1500	Severe	Combined gastrointestinal and bone marrow damage, hypotension	Death within 1 to 2.5 weeks
1500–3000	Severe gastrointestinal damage upper half of range : early transient incapacitation (ETI), gastrointestinal death		Death within 5 to 12 d
3000–4500	Gastrointestinal and cardiovascular damage		Death within 2 to 5 d

lytes. At doses of 12.5 Gy and above, early mortality occurs due to dehydration, and electrolyte imbalance, with death occurring 4 to 5 d after exposure.

REFERENCES

BEIR III. The effects on populations of exposure to low levels of ionizing radiation. Committee on the Biological Effects of Ionizing Radiations, National Research Council. National Academy Press, Washington, DC, 1980.

BEIR IV. Health risks of radon and other internally deposited alpha-emitters. committee on the Biological Effects of Ionizing Radiations, National Research Council. National Academy Press, Washington, DC, 1988.

BEIR V. Health effects of exposure to low levels of ionizing radiation. Committee on the Biological Effects of Ionizing Radiations, National Research Council. National Academy Press, Washington, DC, 1990.

Clarke, R.H. Dose distributions in Western Europe following Chernobyl, in *Radiation and Health. The Biological Effects of Low-Level Exposure to Ionizing Radiation*. R. Jones and R. Southwood, Eds. John Wiley & Sons Ltd., New York, 1987.

Eisenbud, M. *Environmental Radioactivity. From Natural, Industrial, and Military Sources.* 3rd ed. Academic Press. Inc., New York, 1987.

Fetter, S.A. and Tsipis. K. Catastrophic releases of radioactivity, *Sci. Am.* 244 : 41–47, 1981.

Grosch, D.S. Radiations and radioisotopes, in *Introduction to Environmental Toxicology*. F. Guthrie and J. Perry, Eds. Elsevier/North Holland, Inc., New York, 1980,

Moriarity, F. *Ecotoxicology. The Study of Pollutants in Ecosystems*, 2nd ed. Academic Press, San Diego, CA, 1983.

Odum, E.P. Radiation ecology, in *Fundamentals of Ecology*, 3rd ed. W.B. Saunders Company, Philadelphia, 1971.

Robertson, J.B. Toxicology of ionizing radiation, in *A Guide to General Toxicology*. 2nd ed. J. Marquis, Ed. Karger, New York, 1989.

Rust, D.M. Solar flares, proton showers and the space shuttle, *Science*, 216:939–946, 1982.

Sanders, C.L. *Toxicological Aspects of Energy Production*. Battlelle Press, Columbus, OH, 1986.

Appendices

APPENDIX–A

GLOSSARY

Abiotic non-living.

Acid deposition combination of wet deposition from the atmosphere of droplets of sulfuric acid and nitric acid dissolved in rain, sleet, and snow (acid precipitation) and dry deposition from the atmosphere of particles of sulfate and nitrate salts. These acid and salts are formed when water vapour in the air reacts with the air pollutants sulfur dioxide (SO_2) and nitrogen dioxide (NO_2).

Acid mine drainage dissolving and transporting of sulfuric acid and toxic metal compounds from abandoned under ground coal mines to nearby streams and rivers when surface water flows through the mines.

Active solar heating system that uses solar collectors to capture energy from the sun as heat and then uses mechanical devices such as pumps and fans to move the captured heat to a storage system (usually an insulated tank of water or bed of rocks) or throughout dwelling.

Adaptation the process of accommodation to change: the process by which the characteristics of an organism become suited to the environment in which the organism lives.

Adsorption the process by which molecules from a liquid or gaseous phase become concentrated on the surface of a solid.

Aerobic organism organism that require oxygen to live.

Aerobiosis bacterial decomposition in the presence of air.

Aerosol a substance consisting of small particles, typically having diameters that range from 1/100p.m. to 1 p.m.

Air pollutant chemical whose concentration in the atmosphere builds up to the point of causing harm to human beings, other animals, vegetation, or materials such as metal and stone.

Air pollution the deterioration of the quality of air that results from the addition of impurities.

Air quality standards maximum allowable outdoor concentration of air pollutants.

Abedo a measure of the reflectivity of a surface, measured as the ratio of light reflected to light received. A mirror or bright snowy surface has a high albedo, whereas a tough, flat rood surface has a low albedo.

Algae simple one-celled or many celled plants, usually aquatic, capable of carrying on photosynthesis.

Algal bloom population explosion of algae in surface waters resulting from an increase in plant nutrients and phosphates.

Alpha particle positively charged chunks of ionizing radiation consisting of two protons and two neutrons that are spontaneously emitted by the nuclei of some radioisotopes.

Alveoli tiny sacs at the end of the bronchiole tubes in the lungs where oxygen is inhaled, air is transferred to haemoglobin in the blood, and carbon dioxide in the blood is transferred to inhaled air and exhaled.

Ambient air surrounding outdoor air

Anaerobic organism organism that does not require oxygen to survive

Animal manure dung (fecal matter) and urine of animals.

Anthracite hard coal with a low to moderate sulfur content, a very low moisture content (2%), and a high heat content.

Anthropocentric human-centered.

Aquatic pertaining to water.

Aquatic ecosystem any major ecosystem such as river, pond, lake and ocean found in hydrosphere.

Aquifer water-bearing layer of the earth's crust; water in an aquifer is known as groundwater.

Aquifer depletion withdrawal of groundwater from an aquifer faster than if recharged by precipitation.

Arable land land that is capable of being cultivated and supporting agricultural production.

Area strip-mining types of surface mining in which minerals such as coal and phosphate are removed by cutting deep trenches in flat or rolling terrain.

Arid dry, parched with heat.

Artesian well a water well drilled

into a pressurized confined aquifer in which the hydraulic pressure is so great that, the water flows freely up the bore hole to the level of the water in the earth without any need for pumping.

Atmosphere layer of air surrounding the earth's surface.

Atoms extremely small particles that are the basic building blocks of all elements and thus all matter.

Autotroph an organism that obtain its energy from the sun, as opposed to a heterotroph, which is an organism that obtains its energy form the tissue of other organism. Most plants are autotrophs.

Background ionizing radiation ionizing radiation in the environment from naturally radioactive materials and from cosmic rays entering the atmosphere.

Bacteria smallest living organism; with fungi, they comprise the decomposer level of food chains and webs.

Bathyal zone cold, fairly dark zone in an ocean below the euphotic zone which there is some penetration by sunlight but not enough for photosynthesis.

Benthic organism a plant or animal that lives at or near the bottom of a body of water.

Benthic zone bottom of a body of water.

Beta particle swiftly moving electron emitted by the nucleus of a radioisotope.

Biocentric ecosystem-centered.

Biodegradable material that can be broken down into simpler substances (elements and compounds) by bacteria or other decomposers.

Biofuels gas or liquid fuels (such as ethyl alcohol) made from biomass (plants and trees).

Biogas mixture of methane (CH_4) and carbon dioxide (CO_2) gases produced when anaerobic bacteria break down plants and organic waste (such as manure).

Biogeochemical cycle mechanism by which chemicals such as carbon, oxygen, phosphorus, nitrogen, and water are continuously moved through the biosphere to be renewed again and again for use by living organisms.

Biological amplification increase in concentration of certain fat soluble chemicals such as DDT in successively higher tropical levels of a food chain or web.

Biological control use of natural predators, parasites, or disease-causing bacteria and viruses (Pathogens) to regulate the population of a pest species.

Biological oxygen demand (BOD) the amount of dissolved oxygen gas required for bacterial decomposition of organic wastes in water; usually expressed in terms of the parts per million (ppm) of dissolved oxygen consumed over 5 days at 200°C (68°F) and normal atmospheric pressure.

Biomass total dry weight of all living organism in a given area.

Biomass fuel plants and animal matters such as wood that can be burned directly as a source of heat or converted to a move convenient gaseous or liquid biofuel.

Chlorinated hydrocarbon Insecticides class of synthetic, organic compounds containing chlorine, hydrogen, and carbon that are persistent and fat-soluble and can be biologically amplified in food chains and webs. Examples include DDT, aldrin, and dieldrin.

Chlorination addition of chlorine to drinking water or treated sewage plant effluent to kill germs (disinfection)

Chlorofluorocarbons (CFCs) organic molecules containing varying numbers of chlorine, fluorine, carbon, and hydrogen atoms.

Chronic bronchitis lung disorder characterized by persistent inflammation of the bronchi, excessive mucus buildup, recurrent coughing, and throat imitation, it appears to be caused and aggravated by smoking and air pollution.

Clay soil low-porosity soil with a high clay content and little, if any, silt and sand.

Clearcutting methods of timber harvesting in which all trees in a forested area are removed.

Climate average of day to day whether conditions at a given place on earth over a fairly long period, usually 30 years or more. Also includes extremes in weather behaviour during the some period.

Climax ecosystem (Climax community) a relatively stable, self-sustaining stage of ecological succession; mature ecosystem with a diverse array of species and ecological riches, capable of using energy and cycling critical chemicals more efficiently than simpler, immature ecosystem.

Coal a solid, combustible material usually containing from 40% to 98% carbon mixed with varying amounts of water (2% to 50%) and small amounts of nitrogen (0.2% to 1.2%) and sulfur 0.6% to 4%) compounds.

Coal gasification process in which solid coal is converted to liquid hydrocarbon fuel such as methanol or synthetic gasoline.

Coastal wetland land along a coastline that remains flooded with salt water all or part of the year.

Congeneration the production of two useful forms of energy from the same process, in a factory, for example, excess steam produced for industrial processes or space heating is run through turbines to generate electricity, which can be used by the industry or sold to power companies.

Coliform bacteria a normally harmless type of bacteria that resides in the intestinal tract of human beings and other animals and whose presence in water is an indicator that the water be contaminated with other disease causing organism found in untreated human and animal waste.

Coliform bacteria count number of colonies of fecal coliform bacteria present in a 100-milliliter sample of water.

Combustion burning. Any very rapid chemical reaction in which heat and light are produced.

Conservation wise use and careful management of resources, so as to obtain the maximum possible benefits from them for present and future generations. Methods includes preservation, balanced multiple use, reducing unnecessary waste, recycling, reuse, and decreased resources use.

Conservationist people who express their concern for the present and future survival of human being and other species by not wasting and not irreversibly depleting or degrading the biological, physical, and chemical wealth of the word on which all life depends.

Conservation-tillage forming methods of cultivation in which the soil is disturbed very little (minimum till-age farming) or not at all (no-till farming) in order to reduce soil erosion, decrease-labour costs, and save energy.

Consumer organism that relies on other organisms for its food. Generally divided into primary consumers (herbivores), secondary consumers (carnivores), and micro-consumers (decomposers).

Consumption overpopulation situation in the world or a given country or region in which a relatively small number of people use resources of such a high rate and without sufficient population and landuse controls that the resulting population, environmental degradation, and resources depletion can threaten the health and survival of human beings and other species and disrupt the processes that cleanse and replenish the air, water and soil.

Consumption water use water use that results in water being last by evaporation or transpiration or degraded by population so that it is no longer available for reuse in a particular area.

Continental shelf submerged sea floor that slopes gradually from the exposed edge or shore of a continent for a variable distance to a point where a much steeper descent to ocean bottom begins.

Contour farming plowing an planting along rather than up and down the sloped contours of land to reduce soil erosion and conserve water.

Contour strip-farming form of surface mining carried out in hilly or mountains terrain by cutting out a series of shelves or terraces on the side of a hill or mountain and dumping the over-burden from each new terrace onto the one below. Used primarily for cool.

Control rod neutron-absorbing rods that are raised or lowered in the core of a nuclear reactor to control the rate of nuclear fission.

Coral reef shallow area near the coast of a warm tropical or subtropical ocean consisting of calcium-containing material secreted by photosynthesizing red and green algae and small coral animals.

Core (of the earth) central or innermost portion of the earth.

Cosmic rays streams of highly penetrating charged particles composed of portions, alpha particles, and a few heavier nuclei that bombard the earth from outer space.

Cost-benefit analysis technique used to estimate and compare the expected costs or losses associated with a particular project or degree population control with the expected benefits or gains of a given period of time.

Cost effectiveness analysis technique used to determine how a particular goal such as population control, can be achieved for the least cost.

Critical condition in nuclear science, a critical condition said to exist when a chain reaction continues at a steady rate, neither accelerating nor slowing down.

Critical mass the quantity of fissionable material needed to initiate and maintain a nuclear fission chain reaction.

Critical minimum size in general, a critical condition relates to a point at which some property changes very abruptly in response to a small change in some other property of the system. In ecology, a population is said to be reduced to its critical minimum size when its numbers are so few that it is in acute danger of extinction.

Crude oil a gooey liquid mixture of hydrocarbon compounds (90% to 95% of its weight) and small quantities of compounds containing oxygen, sulfur, and nitrogen, which can be extracted from underground deposits and then sent to refineries to be converted to useful materials such as heating oil, diesel fuel, gasoline and far.

Crust (of the earth) solid, outer layer of the earth.

Cultural eutrophication overnourishment of aquatic ecosystem with plant nutrients resulting from human activities such as agriculture, urbanization, and industrial discharge.

DDT *d*ichloro*d*ipheny*t*richloroethane, a chlorinated hydrocarbon that has been widely used as a pesticide.

Deciduous plants plants such as oak and maple that lose all their leaves during part of the year.

Decomposers organisms such as bacteria, mushrooms, and fungi that obtain nutrients by breaking down complex matter in the waste and dead bodies of other organisms into simpler chemicals, most of which are returned to soil and water for reuse by producers.

Deforestation removal of trees from an area without adequate replanting.

Delta build-up deposit of river-borne sediments found near the mouth of a river near the ocean.

Depletion time period required to use up a certain fraction–usually 80%–of the known reserves or estimated resources of a mineral at an assumed rate of use.

Desalinization purification of salt or brackish water by removing the dissolved salts.

Desert biome characterized by very low average annual precipitation (less than 25 centimeters, or 10 inches, a year) and sparse, mostly low vegetation.

Detergent a cleaning agent that acts by binding water molecule: to molecules of grease or other soiling substances.

Desertification conservation of productive grass land, cropland, or forest into desert, usually through combination of overgrazing, prolonged drought, and climate change.

Detritus dead point material, bodies of animals, and fecal matter.

Dioxins family of at least 75 different highly toxic chlorinated hydrocarbon compounds.

Dissolved oxygen (DO) content amount at oxygen gas (O2) dissolved in a given quantity of water of a given temperature and atmospheric pressure, it is usually expressed in parts per million (ppm).

DNA (*deoxyribonucleic acid*) large molecules found in the cells that carry genetic information.

Dredge spoils materials scraped from the bottoms of the harbours and rivers to maintain shipping channels.

Dredging surface mining of seabeds and streambeds, primarily for sand and gravel.

Drip irrigation methods of irrigation in which small pipes deliver water to plant roots.

Drought prolonged period of dry weather.

Dry farming cultivation of agricultural crops without the use of irrigation.

Ecological efficiency flood chain efficiency) the present transfer of useful energy from one tropical level to the next higher tropic level in a food chain.

Ecological equivalents species that occupy the some or similar ecological niches in similar or ecosystem located in different parts of the world. For example, cattle in North America and Kangaroos in Australia are both grassland grazers; hence, they are ecological equivalents.

Ecological description of all the physical, chemical, and biological factors that a species needs to survive, stay healthy, and reproduce in an ecosystem.

Ecological succession process in which communities of plant and animal species are replaced in a particular area over time by a series of different and usually more complex communities.

Ecology study of the interactions of living organisms with each other and with their environment: study of the structure and function of nature.

Ecosystem self-regulating natural community of plants and animals interacting with one another and with their non-living environment.

Ectoparasite that lives outside its host organism.

Effluent any substance, particularly a liquid, that enters the environment from a point source. Generally refers to wastewater from sewage treatment ofindustrial plant.

Electromagnetic radiation radiant energy that can move through a vacuum or through space as waves of oscillating electric and magnetic fields.

Electromagnetic spectrum span of a electromagnetic energy ranging from short-wavelength gamma waves to long-wave-length radio waves.

Emergency core cooling system system designed to prevent meltdown if the core of a nuclear reactor overheats by instantaneous flooding of the core with large amounts of water.

Emission discharge of one or more gases or liquids into the environment.

Emission standard maximum amount of a pollutant that is permitted by the federal government to be discharged into the air or a body of water from a point source.

Emphysema lung disease in which the alveoli enlarge, fuse together, and lose their elasticity, thus impairing the transfer of oxygen to the blood.

Endangered species a wild species having so few individual survivors that it could soon become extinct in all or most of its natural range.

Endoparasite parasite that lives inside its host organism.

Energy ability to do work or produce change by pushing or pulling some form of matter or to cause a heat transfer between two objects of different temperature.

Energy conservation reduction or elimination of unnecessary energy use and waste.

Energy crisis a shortage of a catastrophic price increase for one or more forms of useful energy, or a situation in which energy use is so great that the resulting pollution and environmental degradation threaten human health and welfare.

Energy efficiency the percentage of the total energy input that does useful work and is not converted into low quality, essentially useful, law-temperature heat in an energy conservation system or process.

Energy flow pyramid diagram representing the loss or degradation of useful energy of each step in a food chain. About 80% to 90% of the energy in each transfer is lost as waste heat, and resulting shape of the energy levels is pyramidal.

Energy quality ability of a form of energy to do useful work.

Enhanced oil recovery removal of some of the heavy oil remaining in an oil well after primary and secondary recovery by methods such as pumping in steam or igniting the oil to increase its flow rate so that it can be pumped to the surface.

Entropy a measure of randomness or disorder.

Environment oil of the external conditions that affect an organism or other specified system during its lifetime, everything outside of a specified system.

Environmental degradation depletion or destruction of some renewable resources by using it at a faster rate than it is naturally replenished.

Environmental resistance all the limiting factors that act together to regulate the maximum allowable size, or carrying capacity, of a population.

Epilimnion upper layer of warm water with high levels of dissolved oxygen in a stratified lake.

Erosion removal of soil by flowing water or wind.

Estuarine zone area near coastline that consists of estuaries and coastal saltwater wetlands and that extends out to the edge of the continental shelf.

Estuary thin zone along a coastline where fresh water form rivers mixes with salty ocean water.

Euphotic zone surface layer of an ocean, lake or other body of water through which there is sufficient sunlight for photosynthesis.

Eutrophic lake lake with a large or excessive supply of plant nutrients (mostly nitrates and phosphates).

Eutrophication natural process in which lakes receive inputs of plant nutrients (mostly nitrates and phosphates) as a result of natural erosion and runoff from the surrounding land basin.

Evaporation change of a liquid into vapour.

Evapotranspiration combination of evaporation and transpiration of liquid water in plant tissue and in the soil to water vapour in the atmosphere.

Extinction complete disappearance of on entire species.

Fertilizer substance that makes the land or soil capable of producing more vegetation or crops.

First law of ecology when human beings interfere with or modify on ecosystem, there are always numerous short and long-term effects, many of which are unpredictable.

Fissionable isotopes isotopes that are capable of undergoing nuclear fission.

Floodplain land along a river that

is subject to periodic flooding when the river overflows its banks.

Fly ash small, solid particles of ash and soot generated when coal, oil or waste materials are burned.

Food additive chemical deliberately added to a food, usually to enhance its colour, flavour, shelf life, or nutritional characteristics.

Food chain sequence of transfers of energy in the form of food from organisms in one trophic level of organisms in another tropic level when one organism eats or decomposes another.

Food web complex network of many interconnected food chains and feeding interactions.

Forest region with sufficient average annual precipitation of 75 centimeters (30 inches) or more to support various species of trees and smaller forms of vegetation.

Fossil fuel buried deposits of decoyed plants and animals that have been converted to crude oil, coal natural gas, or heavy oils by exposure to heat and pressure in the earth's crust over hundreds of millions of years.

Fungicide substance or mixture of substances used to prevent or kill fungi.

Fungus simple or complex organism without chlorophyll. The simpler forms are unicellular, the higher forms have branched filaments and complicated life cycle. Examples are moulds, yeasts, and mushrooms.

Gamma rays high-energy electromagnetic waves emitted by the nuclei of some radioisotopes.

Gasohol vehicle fuel consisting of a mixture of gasoline and ethyl or methyl alcohol that typical contains 10% to 23% by volume alcohol.

Gene pool total genetic information possessed by a given reproduction population.

Genetic adaptation changes in the genetic make-up of organism or a species that allows the species to reproduce and gain a competitive advantage under changed environmental conditions.

Genetic damage damage by radiation or chemicals to reproductive cells, resulting in mutations that can be passed on to the future generations in the form of fetal and infant deaths and physical and mental disabilities.

Geothermal energy heat transferred form the earth's intensely hot molten core to underground deposits of dry steam (steam with no water droplets). Wet steam (a mixture of steam and water droplets), hot water, or rocks lying relatively close to earth's surface.

Grassland biome found in regions where moderate average precipitation, ranging from 25 to 75 centimeters (10 to 30 inches) a year, is enough to allow grass to prosper but not enough to support large stands of trees.

Greenhouse gases gases present in the earth's atmosphere that cause the greenhouse effect.

Green manure fresh or still-growing green vegetation plowed into the soil to increase the organic matter and humus available to support crop growth.

Green revolution popular term for the introduction of scientifically bred or selected varieties of a grain (rice, wheat, maize) that with high enough inputs of fertilizer and water can give greatly increased yields per area of land planted.

Groundwater water that sinks into the soil, where it may be stored for ling times in slowly flowing and slowly renewed underground reservoirs known as aquifers.

Gully reclamation using small dams of manure and straw, earth, stone, or concrete to collect silt and gradually fill in channels of eroded soil.

Habitat place or type of place where an organism or community of organisms naturally or normally thrives.

Half-life length of time taken for half the atoms in a given amount of a radioactive substance to emit one or more forms of ionizing radiation and, in the process, change into another non-radioactive or radioactive isotope.

Hazard something that can cause injury, disease, death, economic loss, or environmental deterioration.

Hazardous waste discarded solid, liquid, or gaseous material that may pose a substantial threat to human health or the environment when managed improperly.

Heat form of kinetic energy that flows from one body to anther as a result of a temperature difference between the two bodies.

Heavy oil black, high-sulfur, thick oil found in deposits of crude oil, for sands, and oil shale.

Heavy water water (D_2O) in which oil the hydrogen atoms have bene replaced by deuterium (D).

Herbicide chemical that injures or kills plant life by interfering with normal growth.

Herbivore organism that feeds on plants.

High-grade ore an are that contains a relatively large concentration of a desired metallic element.

High-quality energy energy that is concentrated and has great ability to perform useful work, Example include high temperature heat and the energy in electricity, cool, oil, gasoline, sunlight, and nuclei of uranium-235.

Host plant or animal fed on by a parasite.

Humus complex mixture of partially decomposed, water-insoluble material found in the topsoil layer; it helps retain water and water-soluble nutrients so they can be taken up by plant roots.

Hydrocarbons class of organic compounds containing carbon (C) and hydrogen (H).

Hydrologic cycle biogeochemical cycle that moves and recycles water in various forms through the biosphere.

Hydropower electrical energy produced by falling water.

Hydrosphere region that includes oil the earth's moisture as liquid water (Oceans, smaller bodies of fresh water, and underground aquifers), frozen water (polar ice caps, floating ice, and frozen upper layer of soil known as permafrost), and small amounts of water vapour in the earth's atmosphere.

Incineration controlled process by which combustible solid or liquid wastes are burned and changed into gases.

Industrialized agriculture supplementation of solar energy with large amounts of energy derived from fossil fuels (especially oil and natural gas) to produce large quantities of crops and livestock for sale within the country where it is grown and to other countries.

Industrial smog air pollution, primarily from sulfur dioxide and suspended particulate matter, produced by the burning of coal and oil by industries and in power plants.

Inland wetland land such as a swamp, marsh, or bog found inland that remains flooded all or part of the year with fresh water.

Inorganic compounds substances that consists of chemical combinations of two or more elements other than those used to form organic compounds.

Insecticide substance or mixture of substance intended to prevent, destroy, or repel insects.

Ionizing radiation fast-moving alpha or beta particles or high-energy electromagnetic radiation emitted by radioisotopes that have enough energy to dislodge one or more electrons from atoms it hits to form changed ions, which can react with and damage living tissue.

Ions atoms or groups of atoms with one or more net positive (+) or negative(-) electric charges.

Isotopes two or more forms of a chemical element that have the same number of protons but different mass numbers or numbers or neutrons in their nuclei.

Kerogen solid, waxy mixture of hydrocarbons that is intimately mixed with a fine-grained sedimentary rock. When the rock is heated to high temperature, the kerogen is vaporized and much of the vapour can be condensed to yield shale oil, which can be refined to give petroleum-like products.

Kilocalories (kcal) unit of energy equal to 1,000 calories.

Kilowatt (kw) unit of electrical power equal to 1,000 watts.

Kinetic energy energy that matter has because of its motion and mass.

Kwashiorkor nutritional deficiency (malnutrition) disease that occurs in infant's and very young children when they are weaned from mother's milk to starchy diet that is relatively high in calories but low in protein.

Lake large natural body of standing fresh water formed when water from precipitation, land runoff, or groundwater flow files depressions in the earth created by glaciation, earthquakes, volcanic activity, and crashes of giant meteorites.

Landfarming spreading and mixing of hazardous or other solid or liquid wastes with surface soil to allow biodegradation to less hazardous or nonhazardous materials.

Landfill land waste disposal site that is located without regard to possible pollution or groundwater and surface water resulting from runoff and leaching, waste is converted intermittently with a layer of earth to reduce scavenger, aesthetic, disease, and air pollution problems.

Land-use planning process for deciding that best use of each parcel of land in an area.

Laterite soil found in some tropical areas in which an insoluble concentration of such metals as iron and aluminium is present; soil fertility is generally poor.

Low of conservation of matter in any ordinary physical or chemical change, matter is neither created nor destroyed but merely changed from one form to another.

Law of tolerance the existence, abundance, and distribution of a species are determined by the levels of one or more physical or chemical factors fall above or below the levels tolerated by the species.

Leaching process in which various soil components found in upper layers are dissolved and carried to lower layers and in some cases to groundwater.

Life-cycle cost initial cost plus lifetime operating cost.

Life expectancy average number of years a newborn can be expected to live.

Lignite form of coal with a low heat content and usually a low sulfur content.

Limiting factor factor such as temperature, light, water, or a chemical that limits the existence, growth, abundance, or distribution of an organism.

Limiting factor principle the single physical or chemical factor that is most deficient in an ecosystem determines the presence or absence and population size of a particular species.

Limnetic zone open-water surface layer of a lake through which there is sufficient sunlight for photosynthesis.

Liquefied natural gas (LNG) natural gas which is converted to liquid form by cooling to a very low temperature and then transported by sea in specially designed refrigerated tanker ships.

Liquefied petroleum gas (LPG) the mixture of liquefied propane and butane gas removed from a deposit of natural gas.

Lithosphere region of soil and rock consisting of the earth's upper surface or crust and the upper portion of the mantle of partially molten rock beneath this crust.

Littoral zone shallow waters near the shore of a body of water.

Loams medium-porosity soils consisting of almost equal amounts of sand and silt and somewhat less clay; considered to be best soil for growing crops.

Low-grade ore an are that contains a relatively low concentration of a desired metallic element.

Low-quality energy form of energy such as law-temperature heat that is dispersed or diluted and has little ability to do useful work.

Lung cancer abnormal, accelerated growth of cells in the nocuous membranes of the bronchial passages.

Magma molten rock material within the earth's core.

Malnutrition conditions in which quality of diet is inadequate and an individual's minimum daily requirement for proteins, fats, vitamins, minerals, and other specific nutrients necessary for good health are not met.

Marasmus nutritional deficiency disease that results from a diet low in both calories and protein.

Mass number sum of the number of neutrons and the number of protons in the nucleus of an atom. It is a measure of the approximate mass of that atom.

Matter anything that has mass and occupies space.

Megawatt (Mw) unit of electrical power equal to 1,000 kilowatts, or 1 million watts.

Mesotrophic lake lake with a moderate supply of plant nutrients.

Metabolic reserve lower half of rangeland grass plants that can grow back as long as it is not consumed by herbivores.

Metallic mineral inorganic substances found in the earth's crust that contains a useful metallic element such as aluminium, iron, or uranium.

Metastasis release of malignant (cancerous) cells from a tumour into other parts of the body.

Microorganism generally, any living thing of microscopic size; example include bacteria, yeasts, simple fungi, some algae, slime molds, and protozoans.

Mineral an inorganic substance (element or compound) occurring naturally in the earth's crust.

Mineral deposit any naturally occurring concentration in the lithosphere of a free element or compound in the solid form.

Mineral resources nonrenewable chemical element or compound usually in solid that is used by people. Mineral resources are classified as metallic (such as fossil fuels, sand, and slat).

Minimum-tillage farming planting crops by disturbing the soil as little as possible and keeping crop residues and litter on the ground instead of turning them under by plowing.

Molecule chemical combination of two or more atoms of the same chemical element (such as O_2) or different chemical elements (such as H_2O).

Monoculture cultivation of a single crop (such as maize or cotton) to the exclusion of other crops on a piece of land.

Municipal waste combined residential and commercial waste materials generated in a given municipal area.

Mutagen any substance capable of increasing the rate of genetic mutation of living organisms.

Mutation inheritable changes in the DNA molecules found in genes as a result of exposure to various environmental factors such as radiation and certain chemicals or during cell division (asexual reproduction) and when a sperm and egg cell fuse (sexual reproduction).

Mutualism interaction between species in which oil species involved are benefited.

National ambient air quality standard (NAAQS) federal standard that specifies the maximum allowable level, averaged over a specific time period, for a certain pollutant in outdoor (ambient) air.

Natural change rate how fast a population is growing or decreasing per year; usually expressed with a percentage and obtained by subtracting the death rate from the birth rate dividing the result by 10.

Natural gas underground deposits of gases consisting of 50% to 90% methane (CH_4) and small amounts of heavier gaseous hydrocarbon compounds such as propane (C_3H_6) and butane (C_4H_{10}).

Natural radioactivity a nuclear change in which unstable nuclei of atoms spontaneously shoot out "chunks" of mass energy, or both at a fixed rate.

Natural resource anything obtained from physical environment to meet human needs.

Natural selection mechanism for evolutionary change in which individual organisms in a single population die off over time because they cannot tolerate a new stress and are replaced by individuals whose genetic traits allow them to cope with the stress and reproduce successfully to pass these adoptive traits on to their offspring.

Neo-Malthusians people who

believe that if present friends continue, the world will become more crowded and more polluted, leading to greater polluted, leading to greater political and economic instability and increasing the threat of nuclear war of the rich get richer and poor get poorer.

Neritic zone relatively warm, nutrient rich, shallow portion of the ocean that extends from the high-tide mark on land to the edge of the continental shelf.

Net useful energy total useful energy available from an energy resource or energy system minus the useful energy used, lost and wasted in finding, processing, concentrating, and transporting it to a user.

Neutron (N) elementary particles present in the nuclei of all atoms (expect hydrogen-1). It has relative mass of 1 and no electric charge.

Nitrogen cycle biogeochemical cycle in which nitrogen is converted into various forms and transported through the biosphere.

Nitrogen fixation process in which bacteria and other soil microorganisms convert atmospheric nitrogen into nitrates, which become available to growing plants.

Nonbiodegradable pollutant material, such as toxic mercury and lead compounds, that is not broken down to an acceptable level or form in the environment by natural processes.

Nonmetallic mineral inorganic substance found in the earth's crust that contains useful nonmetallic compounds such as those in sand, stone, and nitrate and phosphate salts used as commercial fertilizers.

Nonpoint source source of pollution in which wastes are not released at one specific, identifiable point but from a number of points that are spread out and difficult to identify and control.

Nuclear change process in which the isotope of an element changes into one or more different isotopes by altering the number of protons, neutrons, or both in its nucleus.

Nuclear energy energy released when atomic nuclei undergo fission or fusion.

Nuclear fission nuclear change in which the nuclei of certain heavy isotopes with large numbers such as uranium-235 are split a part into two lighter nuclei when struck by slow-or fast moving neutrons and a substantial amount of energy.

Nuclear fusion nuclear change in which two nuclei of light elements such as hydrogen are forced together at high temperatures of 100 million to 1 billion C until they fuse to form a heavier nucleus with the release of a substantial amount of energy.

Nuclear winter effect significant drop in atmospheric temperature and degree of light penetration to the earth's surface as a result of massive amounts of smoke, soot, dust, and other debris lifted into the atmosphere by a limited nuclear war.

Nucleus the extremely tiny centre of an atom, which contains one or more positively charged protons and in most cases one or more neutrons with no electrical charge. The nucleus contains most of an atom's mass.

Nutrient element or compound needed for the survival, growth, and reproduction of a plant or animal.

Oil shale underground information of a fine-grained rock that contains varying amounts of a solid, waxy mixture of hydrocarbon compounds known as kerogen. When the rock is heated to high temperatures, the kerogen is converted to a vapour that can be condensed to form a slowflowing heavy oil called shale oil.

Oligotrophic lake a lake with a low supply of plant nutrients.

Omnivore organism such as a pig, rat, cockroach, or human being that can use both plants and other animals as food sources.

Open dump land disposal-site where wastes are deposited and left uncovered with little or no regard for control of scavenger, disease, air pollution, or water pollution problems or for aesthetics.

Open-pit surface mining surface mining of materials (prima stone, sand, gravel, iron, and copper) that creates a large pit.

Open sea the part of an ocean beyond the continental shelf.

Ore mineral deposit containing a high enough concentration of at least one metallic element to permit the metal to be extracted and sold at a profit.

Organic farming methods of producing crops and livestock naturally by using organic fertilizer (manure, legumes, compost, crop residues), crop rotation, and natural pest control (bugs that eat harmful bugs, and environmental controls such as crop rotation instead of using commercial fertilizer and synthetic pesticides and herbicides.

Organic fertilizer organic material such as animal manure, green manner, and compost applied to cropland as a source of plant nutrients.

Organism any form of life.

Organophosphates diverse group of nonpersistent synthetic chemical insecticides that act chiefly by breaking down nerve and muscle responses; examples are parathion and malathion.

Output pollution control method for reducing the level of pollution once a pollutant has entered the environment.

Overburden layer of soil and rock overlying a mineral deposit that is removed during surface mining.

Overfishing harvesting so many fish, especially immature individuals, of a species that not enough breeding stock is left for adequate annual renewal.

Overgrazing excessive grazing of rangeland by livestock to the point of which it cannot be renewed or is renewed slowly because of damage to the root system.

Overnutrition diet so high in calories, saturated (animal) fats, slat, sugar, and processed foods, and so low in vegetables and fruits that the consumer runs high risks of diabetes, hypertension, heart disease, and other health hazards.

Overpopulation impairment of the life-support system in a country, a region, or the world when its people use nonrenewable and renewable resources to such an extent that the resources base is degraded or depleted and air, water, and soil are severely polluted.

Oxygen cycle biogeochemical cycle in which oxygen is converted into various forms and transported through the biosphere.

Oxygen-demanding wastes organic water pollutants that are usually degraded be aerobic (oxygen-consuming) bacteria if there is sufficient dissolved oxygen (DO) in the water.

Ozone layer layer of gaseous ozone (O_3) in the upper atmosphere that protects life on earth by filtering out harmful ultraviolet radiation form the sun.

Parasite primary, secondary, or higher consumer that feeds on a plant or animal, known as a host, over and extended period of time.

Paratransit transit system such as carpools, vanpools, jitneys, and dial-a-ride system that carry a relatively small number of passengers per vehicular unit.

Particulate matter solid particles or liquid droplets suspended or carried in the air.

Parts per billion (ppb) numbers of part of a chemical found in one billion parts of a particular gas, liquid, or solid mixture.

Parts per million (ppm) number of parts of a chemical found in one million parts of a particular gas, liquid, or solid mixture.

Parts per trillion (ppt) numbers of parts of a chemical found in one trillion parts of a particular gas, liquid, or solid mixture.

Pathogen organism that produces disease.

PCBs (*polychlorinatedblphenyls*) mixture of at least 50 widely used organic compounds containing chlorine that can be biologically magnified in food chains and food webs with unknown effects.

Peat a fuel with a low heat content and high moisture content (70% to 95%) that is the first step in the formation of various types of coal.

Permafrost water permanently frozen year-round in thick underground layers of soil found in tundra.

Peroxyacyl nitrates group of chemicals (photochemical oxidants) also known as PANs, found in photochemical smog.

Perpetual resource a resource such as solar energy that comes from an essentially inexhaustible source and thus will always be available on human time scale regardless of whether or how it is used.

Pest unwanted organism that directly or indirectly interferes with human activities.

Pesticide any chemical designed to kill weeds, insects, fungi, rodents, and other organisms that humans consider to be undesirable.

Pesticide treadmill station in which the costs of using pesticides increase while their effectiveness decreases, primarily as a result of genetic resistance to the chemicals by target organisms.

Petrochemicals chemicals obtained by refining (Distilling) crude oil that are used as raw materials in the manufacture of most industrial chemicals, fertilizers, pesticides, plastics, synthetic fibres, paints, medicines, and numerous other products.

pH numeric value that indicates the relative acidity or alkalinity of a substance on a scale of 0 to 14, with the neutral point at 7.0. Acid solutions have pH values lower than 7.0 and basic solutions have pH values greater than 7.

Phosphorus cycle biogeochemical cycle in which phosphorus is converted into various chemical forms and transported through the biosphere.

Photosynthesis complex process that occurs in the cells of green plants whereby radiant energy from the sub is used to combine carbon dioxide (CO_2) and water H_2O) to produce oxygen (O_2) and simple sugar or food molecules, such as glucose ($C_6H_{12}O_6$).

Photovoltaic cell (solar cell) device in which radiant (solar) energy is converted directly into electrical energy.

Phytoplankton free-floating, mostly microscopic aquatic plants.

Plankton microscopic floating plant and animal organisms of lakes, rivers, and oceans.

Point source source of pollution that involves discharge of pollutants from an identification point, such as a smokestack or sewage treatment plant.

Pollution a change in the physical, chemical, or biological characteristics of the air, water, or soil that can affect the health, survival, or activities of human beings or other living organisms in a harmful way.

Potential energy energy stored in an object as a result of its position or the position of its parts.

Precipitation water in the form of rain, sleet, hail, and snow that falls from the atmosphere onto land and bodies of water.

Predator organism that captures and feeds on parts or all of an organism of another species (the prey).

Prey organism that is captured and serves as a sources of food for an organism of another species (the predator).

Primary succession sequential development of communities in a bare or soilless area that has never been or occupied by a community of organisms.

Prime reproductive age years between ages 20 to 29 during which most women have most of their children.

Producer organism that uses solar energy (green plant) or chemical energy (some bacteria) to manufacture its own organic substances (food) from inorganic nutrients.

Profundal zone deepwater region of a lake, which is not penetrated by sunlight.

Proton (p) positively charged particle found in the nuclei of all atoms. Each proton has a relative mass of 1 and a single positive charge.

Pyramid of biomass diagram representing the biomass, or total dry weight of all living organisms, that can be supported at each trophic level in a food chain.

Pyramid of numbers diagram representing the number of organisms of a particular type that can be supported of each trophic level from a given input of solar energy in food chains and food webs.

Pyrolysis high-temperature decomposition of material in the absence of oxygen.

Radiation propagation of energy through matter and space in the form of fast-moving particles (particulate radiation or waves (electromagnetic radiation).

Radioactive fallout radioactive dirt and debris that falls back to earth after being released into the atmosphere as a result of the detonation of nuclear weapon or an accident at a nuclear power plant or other facility handling radioactive materials.

Radioactive waste end products of nuclear power plants, research, medicine, weapons production, or other processes involving radioisotopes.

Radioisotope isotope of an atom whose unstable nuclei spontaneously emit fast moving particles (such as alpha or beta particles), high-energy electromagnetic radiation in the form of gamma rays, or both to form nonradioactive or radioactive isotopes of a different kind.

Rangeland land on which the vegetation is predominantly grasses, grasslike plants, or shrubs such as sagebrush and that is capable of providing forage for grazing or browsing animals.

Range of tolerance range or span of chemical and physical conditions that must be maintained for populations of a particular species to stay alive and grow, develop, and function normally.

Rapidly biodegradable pollutant

substance such as human sewage that can be rapidly broken down in the environment into an acceptable level or form by natural processes.

Recharge area area in which an aquifer is replenished with water by the downward percolation through soil and rock.

Recharging replenishment of water in an aquifer.

Recycling collecting and remitting or repossessing a resource so it can be used again, as when used glass bottles are collected, melted down, and made into new glass bottles.

Renewable resource resource that can be depleted in the short run if used or contaminated too rapidly but normally will be replaced through natural processes in the long urn.

Reserves identified deposits of a particular resource in known locations that can be extracted profitably at present prices and with current mining technology.

Reservoir large and deep, human created body of standing fresh water often filled behind a dam.

Resource conservation developing and protecting natural resources for the greatest good of the greatest number of people for the longest length of time by reducing unnecessary resource use and waste.

Resource recovery extraction of useful materials or energy from waste materials. This may involve recycling or conversion into different and sometimes unrelated products or uses.

Resource recovery plant centralized facility in which mixed urban solid waste is shredded and automatically separated to recover glass, iron, aluminium, and their valuable materials. The remaining paper, plastics, and other materials, are incinerated to produce steam, hot water, or electricity.

Resources identified and unidentified deposits of a particular mineral that cannot be recovered profitably with present prices and mining technology but may be converted to reserves when prices rise or mining technology improves.

Reuse to use a product again and again in the some, as when returnable glass bottles are washed and refilled.

Risk probability that something undesirable will happen from deliberate or accidental exposure to a hazard.

Risk assessment process of determining the short-and long-term adverse consequences to individuals or groups from the use of a particular technology in a particular area.

Risk-benefit analysis estimating the short-and long-term societal benefits and risks involved in using a particular technology; the benefits can be divided by the risks to find a desirability quotient that can be used to help determine whether the technology should be used.

River fairly wide and deep-flowing body of water that usually empties into an ocean.

Rocky shore steep, rock-laden coastline.

Rodenticide substance that can kill rodents.

Ruminant animals animals such as cattle, sheep, goats, and buffalo with four-chambered stomachs that digest cellulose.

Runoff surface water entering rivers, freshwater lakes, or reservoirs from land surfaces.

Salinity amount of dissolved salts) especially sodium chloride) in a given volume of water.

Salinization accumulation of salts in soils that can eventually render the soil incapable of supporting plant growth.

Saltwater intrusion movement of salt water into freshwater aquifers in coastal areas as groundwater is withdrawn faster than it is recharged by precipitation.

Sandy soil highly porous soil containing a large amount of sand and little, if any, silt and day.

Sanitary landfill land waste disposal site located to minimize water pollution from runoff and leaching waste is spread in thin layers, compacted and covered with a fresh layer of soil each day.

Secondary air pollutant harmful chemical formed in the atmosphere through a chemical reaction among air components.

Secondary oil recovery injection of water to force some of the remaining crude oil from a well after primary recovery. Usually primary and secondary recover remove about one-third of the crude oil in a well.

Secondary succession Sequential development of communities in an area in which natural vegetation has been removed or destroyed, but the soil or bottom sediment is not destroyed.

Secured landfill a land site for the storage of hazardous solid and liquid wastes that are normally placed in containers and buried in a restricted-access area that is continually monitored. Such landfills are located above geologic strata that are supposed to prevent the leaching of wastes into ground-water.

Sediment soil particles, sand, and minerals washed from the land into aquatic systems as a result of natural and human activities.

Seed-tree cutting removal of nearly all trees on a site in one cut, with a few of the better commercially valuable trees left uniformly distributed as a source of seed to regenerate the forest.

Selective cutting cutting of intermediate-aged or mature of diseased trees in an uneven-aged forest stand either singly or in small groups to encourage younger trees to grow and to produce an uneven-aged stand with trees of different species, ages, and sizes.

Septic tank underground receptacle for wastewater from a home in rural and suburban areas. The bacteria in the sewage decompose the organic wastes and the sludge settles to the bottom of the tank. The effluent flows out of the tank into the ground through a field of drain pipes.

Shale oil a slow-flowing, dark brown, heavy oil obtained when kerogen in shale oil rock is vaporized at high temperatures and then condensed. Shale oil can be refined yield petroleum products.

Shelterwood cutting removal of oil mature trees in an area in a series of cuts over one or more decodes.

Shifting cultivation clearing and planting a plot of ground in a forest for two to five years, until no further cultivation is worthwhile because of a reduction of soil fertility or invasion by a dense growth of vegetation, and then clearing a new plot to grow crops by slash-and-burn cultivation.

Silviculture cultivation and management of forest to produce a renewable supply of timber.

Slash-and-burn cultivation in many tropical areas, the practice of cleaning a patch of forest, leaving the cut vegetation on the ground to dry, burning the dried residue to add nutrients to the soil, and planting crops, ideally the patch is abandoned after two or five years of cultivation to prevent depletion of soil fertility.

Slowly biodegradable pollutant a pollutant such as DDT that is broken down in the environment by natural processes to an acceptable level of form at a relatively slow rate.

Soil complex mixture of inorganic minerals (mostly clay, silt, and sand), decaying organic matter, water, air, and living organisms.

Soil conservation methods used to reduce soil erosion and prevent depletion of soil nutrients.

Soil erosion movement of soil components, especially topsoil, from one place to another, usually by exposure to wind, flowing water, or both.

Soil horizons horizontal layers that make up a particular type of soil.

Soil porosity numbers of pores and the average distances between pores in a given sample of soil.

Solar cell device for collecting radiant energy from the sun and converting it into heat.

Solar energy direct radiant energy from the sun plus indirect forms of energy such as wind, falling or flowing water (hydropower), ocean thermal gradients, and biomass- that are produced when solar energy interacts with the earth.

Solar fumace device for collecting and concentrating radiant energy from the sun to reach temperatures high enough to melt metals and carry out other high-temperature operations.

Solar pond relatively small body of fresh water or salt water in which stored solar energy can be extracted as a result of the temperature difference between the surface layer and bottom layer.

Solid waste any unwanted or discarded material that is not a liquid or a gas.

Speciation splitting of a single species over thousands to millions of years into two or more different species in response to new environmental conditions.

Species all organisms of the same kind; a group of plant or animals potentially capable of breeding with other members of its group but normally not with organisms outside its group.

Species diversity number of different species and their relative abundances in a given area.

Sport hunting killing of animals for recreation.

Stability ability of a living system to withstand or recover from externally imposed changes or stresses.

Strategic materials fuel and non-fuel minerals vital to the industry and defence of a country. Ideally supplies are stock-piled to cushion against supply interruptions and sharp price increased.

Strip cropping planting regular crops and close-growing plants such as hay or nitrogen-fixing legumes in alternating rows or bands.

Subatomic particles extremely small particles such as electrons, protons, and neutrons that make up the internal structure of atoms.

Subbituminous coal form of coal with a low heat content and usually a low sulfur content.

Subsidence sinking down of part of the earth's crust resulting form underground excavation - such as a coal mine - or removal of groundwater.

Subsistence agriculture supplementation of solar energy with energy from human labour and draft animals to produce enough food to feed one's self and family members; occasionally some may be left over to sell or put aside for hard times.

Subsistence hunting killing of animals to provide enough food and other materials for survival.

Subsurface mining underground extraction of a metal ore or fuel resource such as coal.

Succulent plants plants such as cacti that store water and produce the food they need in the thick, fleshy tissue of their green stems and branches.

Surface mining the process of removing the over burden of topsoil, subsoil, and other strata to permit the extraction of underlying mineral deposits.

Surface water precipitation that does not infiltrate into the ground or return to the atmosphere and becomes runoff that flows into nearby streams rivers, lakes, wetlands, and reservoirs.

Surroundings (environment) everything outside a specified system or collection of matter.

Synergism interaction in which the total effect is greater than the sum of two effects taken independently.

Synergistic effect result of the interaction of two or more substances or factors that cause a net effect greater than that expected from adding together their independent effects.

Synfuels synthetic gaseous and liquid fuels produced from coal or sources other than natural gas or crude oil.

Synthetic natural gas (SNG) gaseous fuel containing mostly methane that is produced from solid coal.

System any collection of matter under study.

Tailings rock and other waste materials removed as impurities when minerals are mined and mineral deposits are processed. These materials are usually piled on the ground or dumped into ponds.

Tar sands swamplike deposits of a mixture of fine clay, sand, water, and variable amounts of a tarlike heavy oil known as bitumen. The bitumen or heavy oil can be extracted from the tar sand by heating and purified and upgraded to synthetic crude oil.

Teratogen substance that, if ingested by a pregnant female, courses malformation of the developing fetus.

Terracing planting crops on a long, steep slope that has been converted into a series of broad, nearly level terraces with short vertical drops from one to another, and following the slope of the land in order to retain water and reduce soil erosion.

Terrestrial pertaining to the land.

Thermal enrichment beneficial effects in an aquatic ecosystem as a result of a rise in water temperature.

Thermal shock harmful ecological effects in an aquatic ecosystem as a result of a sharp rise or drop in water temperature.

Thermocline fairly thin transition zone in a lake that separates an upper warmer zone (epilimnion) from a lower colder zone (hypolimnion).

Threatened species a wild species that is still abundant in its natural range but is considered likely to become endangered within the foreseeable future because of a decline in numbers.

Threshold effect a harmful or fatal effect that does not occur until the level of a particular physical or chemical factor exceeds the limit of tolerance of an organism.

Tolerance limit point at and beyond which a chemical or physical condition (such as heat) becomes harmful to a living organism.

Total resources total amount of a particular mineral that exists on earth.

Transpiration transfer of water from exposed parts of plants through leaf pores to the atmosphere.

Trickling filters form of secondary sewage treatment in which aerobic bacteria biodegrade organic wastes in waste-water as it seeps through a large vat filled with crushed stones covered with bacterial growths.

Tritium (T: hydrogen-3) isotope of hydrogen with a nucleus containing one proton and two neutrons, thus having a mass number of 3.

Trophic level oil organisms that consume the same general types of food in a food chain or food web. For example, oil producers belong to the first trophic level and all primary consumers belong to the second trophic level in a food chain or a food web.

Troposphere innermost layer of the atmosphere, which contains about 95% of the earth's air and extends about 8 to 12 kilometers (5 to 7 miles) above the earth's surface.

Unconfined aquifer water bearing layer of the earth's crust when ground-water collects above a layer of relatively impermeable rock or compacted clay.

Undernutrition condition characterized by an insufficient quantity or caloric intake of food to meet an individual's minimum daily energy requirement.

Undiscovered resources potential supplies of a particular mineral resource believed to exist on the basis of broad geological knowledge and theory, although location, quality, and amounts are unknown.

Upwelling area along a steep coastal area where winds blow surface water way from the shore and allow cold, nutrient-rich bottom water to rise to the surface.

Waterlogging saturation of soil with irrigation water so that the water table rises close to the surface.

Water pollution any physical or chemical change in surface water of ground-water that can adversely affect living organisms.

Watershed land area that delivers runoff water, sediment, and dissolved substances to a major river and its tributaries.

Water table tap of the water-saturated portion of an unconfined aquifer.

Watt unit of power, or rate of which electrical work is done.

Wavelength distance between the crest (or trough) of one wave of electromagnetic radiation and that of the next.

Weather process in which bedrock is gradually broken down into small bits and pieces that make up most of a soil's inorganic material as a result of exposure to physical and chemical processes.

Wetland land that remains flooded all or part of the year with fresh or salt water.

Whole-free harvesting use of machines to pull entire trees from the ground and reduce them to small chips.

Wilderness area where the earth and its community of life have not been seriously disturbed by human beings and where human beings are only temporary visitors.

Wilderness species wild animal species that flourish only in relatively undisturbed climax vegetational communities such as large areas of mature forest, tundra, grassland, and desert.

Wildlife all free, undomesticated species of plants and animals on earth.

Wildlife conservation the world wide social movement to bring about the protection, preservation, management, and study of wildlife and wildlife resources.

APPENDIX–B

BIOGEOCHEMICAL CYCLES

The elements are distributed in different amounts in the different environmental segments. The flow of a chemical element between one segment and another i.e. between land mass (the edaphic environment), ocean (the hydrosphere) and the atmosphere can be described by means of a cycle known as a *biogeochemical cycle.* All of these movements might be described as natural since they occurred before man appeared on the Earth. The activities of human population have caused changes in the rates of movements of many chemical substances. Not only there are increase in transfer rates of compounds already present (such as carbon dioxide which is being released by burning carbon-containing fuels), but completely non-chemicals (such as DDT and CFCs) are being distributed on land, sea and into the atmosphere.

Each biogeochemical cycle is a model that describes the movement of a chemical species or element in the biosphere. The vast majority of the material stays in this region i.e. the crust, the ocean and the atmosphere. There appears to be little loss either to the mantle of the Earth or the outer space. The cycles are described using reservoirs, physically well-defined units and transport pathways along which the material is transferred from one *reservoir* to another.

In quantitative biogeochemical cycle, estimates are made of both the amount of an element in each reservoir and the flux between the reservoirs. The *flux* is defined as the amount of material passing along a transport pathway in a fixed period of time usually one year.

If the concentration of an element remains constant in a reservoir this means that there is a balance between the amount of element entering the reservoir and the amount leaving. A steady state has been reached. Most narural cycles (those not influenced by activities of man) appear on a global scale to be in a steady state.

If it is known that a system is in a steady state, it is possible to determine the *residence time* for individual elements in particular reservoirs.

$$\text{Residence Time} = \frac{\text{Amount of element in a reservoir}}{\text{rate of addition (or removal) of the element}}$$

For example, if the amount of sodium dissolved in the ocean is 15 × 10^{18} kg, and the amount added each year is 100 × 10^{9} kg, then the residence time will be 150 million years. Most of the elements have long to residence times in the ocean. Residence times in the crust are much longer, and the residence times in the atmosphere are shorter, reflecting the mobility of the systems.

Biogeochemical cycle, of three elements—carbon, nitrogen and sulphur—are considered in the following text.

The Carbon Cycle

Fig. B.1 presents a simplified form of the global carbon cycle. In this figure, the pools of living and dead organic matter on land and in the oceans are assumed to be in equilibrium. Half the CO_2 added to the atmosphere by combustion of fossil fuels is assumed to dissolve in the surface layer of oceans. Build-up of carbon dioxide in the atmosphere is a matter of concern. The combustion of fossil fuels increases the atmospheric CO_2 concentrations by about 0.7 percent per year.

Not all the added CO_2 remains in the atmosphere, however. Some is absorbed by oceans, and some is being taken up in an increase in biomass. An increase in the concentration of CO_2 in the atmosphere should increase the rate of net photosyntheses.

The *producers* (plants) through photosynthesis reduce CO_2 from the atmosphere to organic carbon. This then is passed on to *consumers* (animals and humans) and *decomposers* (fungi and bacteria) and eventually is returned to the atmosphere (as CO_2) through respiration and decomposition. This route is the normal

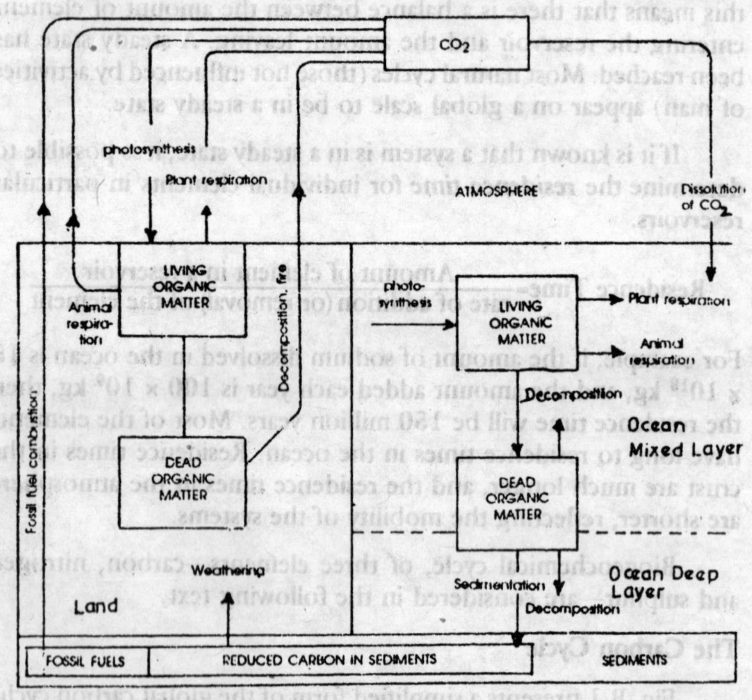

Fig. B.1 : The global carbon cycle

biological activity. Interchange of CO_2 between the oceans (which store more than fifty times as much CO_2 as the atmosphere) in and the atmosphere occurs through diffusion. The ocean reservoir tends to regulate the atmospheric CO_2 concentration.

Significant reserves of bound carbon in the form of inorganic deposits (such as limestone, fossil fuels mainly consisting of coal and petroleum are present in sedimentary of deposits. Combustion of fuels, weathering of rock, and volcanic activity are some of the pathways though bound carbon returns to the atmosphere or oceans in the form of CO_2.

Some other important pathways in the carbon cycle are the

conversion of formaldehyde (HCHO), methane (CH_4), and CO. The main reactions are :

$$2\ H\ CHO \rightarrow CH_4 + CO_2$$

$$2\ HCHO + O_2 \rightarrow 2\ CO + 2\ H_2O$$

The CH_4 and CO may then be oxidized in the soil or in the atmosphere according to the following reactions :

$$2\ CH_4 + 3O_2 \rightarrow 2\ CO + 4\ H_2O \text{ and}$$

$$2\ CO + O_2 \rightarrow 2\ CO_2$$

The Nitrogen Cycle

The nitrogen cycle is chemically the most intricate of all the cycles and probably the least well-understood scientifically. The several chemical form of nitrogen encountered in the nitrogen cycle are given in Table B.2. The crucial role of nitrogen in all proteins makes this element essential to life, but most organisms can assimilate and use nitrogen in specific chemical forms. Only a relatively few species of bacteria and algae can convert gaseous nitrogen, N_2, into more complex compounds that can be used by plants and animals. The main usable forms are NH_3 and NO_3^- of which the latter is required in larger quantity by most plants. The conversion of N_2 to NH_3 and NO_3^- is called nitrogen fixation and the species that can accomplish this conversion are called nitrogen fixing organisms. The continuation of life on the Earth depends absolutely on the activities of these tiny organisms.

Table B.1 : Chemical Forms of Nitrogen

Formula	*Name*	*Oxidation number*	*Comments*
NH_3	Ammonia	–3	Major nutrient form
NH_4+	Ammonium ion	–3	From NH_3 dissolved in water
NH_2+	Amino group	–1	Constituent of protein
N_2	Nitrogen gas	0	Bulk of atmosphere
N_2O	Nitrous oxide	+1	Laughing gas, controls natural ozone cycle
NO	Nitric oxide	+2	Combustion product
NO_2-	Nitrite ion	+3	Link in N cycle
NO_2	Nitrogen dioxide	+4	From NO oxidized in atmosphere
NO_3-	Nitrate ion	+5	Principal nutrient form

The best known of the nitrogen fixing organisms are the bacteria (genus *Rizobium*) associated with the special modules on roots of legumes. In a symbiotic relationship, the bacteria obtain from the plants the energy they need to carry out the nitrogen fixing reaction

$$2N_2 + 6H_2O \rightarrow 4\ NH_3 + 3\ O_2$$

and some of the ammonia, in turn, is made available to the plants for syntheses of amino acids.

$$2NH_3 + 2H_2O + 4\ CO_2 \rightarrow 2\ \underset{\text{(glycine)}}{H_2N\ CH_2\ COOH} + 3\ O_2$$

Although nitrogen fixation by the legume bacteria symbiosis is considered to be responsible for most of the natural nitrogen fixation, the fixation step is also carried out by some bacteria that have looser associations with plants, including the free-living *Azobacter* (aerobic) and clostridium (anaerobic) and by various species of blue green algae.

The amino acids used by all organisms as the building blocks of protein are synthesized by not only from ammonia but also from nitrate produced from ammonia by processes collectively called *nitrification*. The nitrification reactions yield energy, which supports the life processes of the nitrifying bacteria (*Nitrosomonas, Nitrobacter*). The actual synthesis of amino acids is carried out by bacteria, by plants and by animals.

Fixed nitrogen that has been incorporated into organisms returns to the soil in animal wastes and in dead organisms (microbes, plants, animals) or parts of organisms (for example leaves). Animal wastes are rich in urea, $(NH_2)_2CO$, which is the principal product of the metabolism of proteins in many organisms. The proteins of dead organisms are broken down into amino acids and other residues by bacteria and fungi of decay. Urea, amino acids and other breakdown products of protein are then converted into ammonia by yet another group of bacteria. This step is called ammonification.

The mechanism that returns nitrogen tried up as nitrates in soil, ocean and sediments to the atmosphere in its molecular form (N_2) involves another set of bacteria. This group of bacteria is called *denitrifying bacteria*. They convert NO_3^- to N_2O and N_2. The nitrons oxide (N_2O) is reduced to N_2 by further bacterial action or in the atmosphere by photochemical reactions. Another set of denitrification reactions, also carried out by the bacteria, transforms NO_3^- to NO_2^- and NO_2^- to NH_3.

The chemistry of the main steps in the nitrogen cycle assisted by bacteria is given in Table B.2. The part of the nitrogen cycle directly associated with organisms is represented in Fig. B.2. The biological and chemical steps in the global nitrogen cycle are depicted together in Fig. B.3.

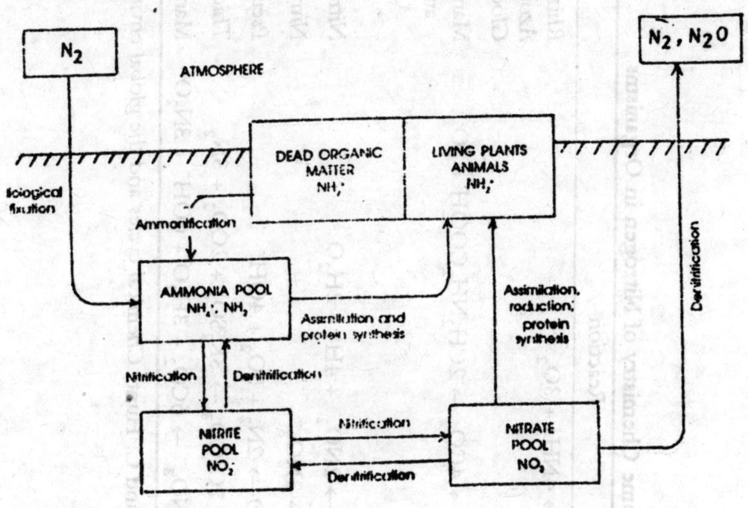

Fig. B.2 : Biological processes in the terrestrial nitrogen cycle

Inorganic Processes : Some transformations in the nitrogen cycle take place without the active participation by the decomposition of organic materials entered the atmosphere by outgassing from the Earth's surface. Being highly soluble in water, ammonia dissolves in atmospheric vapour to form ammonium ion, NH_4^+, which combines with SO_4^{2-} and NO_3^- ions and rain out as $(NH_4)_2SO_4$ and NH_4NO_3.

N_2O, produced by denitrifying bacteria, is the second most abundant form of nitrogen in the atmosphere, after N_2. NO sinks for this gas are known in the tropospheres where it is inert. N_2O is

Table B.2 : Some Chemistry of Nitrogen in Organisms

Step	*Reaction*	*Organism*
Fixation	$2N_2 + 6H_2O \rightarrow 4NH_3 + 3O_2$	*Rhizobium*, *Azotobacter*, *Gloeocapsa*, *Plectonema*
Amino-Acid Synthesis (ammonification is the reverse)	$2NH_3 + 2H_2O + 4CO_2 \rightarrow 2CH_2NH_2COOH + 3O_2$	Many, bacteria and others
Nitrification	$2NH_4^+ + 3O_2 \rightarrow 2NO_2^- + 4H^+ + 2H_2O$	*Nitrosomonas*
Nitrification	$2NO_2^- + O_2 \rightarrow 2NO_3^-$	*Nitrobacter*
Denitrification	$4NO_3^- + 2H_2O \rightarrow 2N_2 + 5O_2 + 4OH^-$	*Pseudomonas*
Denitrification	$5S + 6KNO_3 + 2CaCO_3 \rightarrow 3K_2SO_4 + 2CO_2 + 3N_2$	*Thiobacillus denitrificans*
Denitrification	$C_6H_{12}O_6 + 6\ NO_3^- \rightarrow 6CO_2 + 3H_2O + 6OH^- + 3N_2O$	Many

Sources : R.M. Garrels, F.T. Mackenzie, and C. Hunt, *Chemical cycles and the global environment*; R.C. Burns and R.W.F. Hardy, *Nitrogen fixation*.

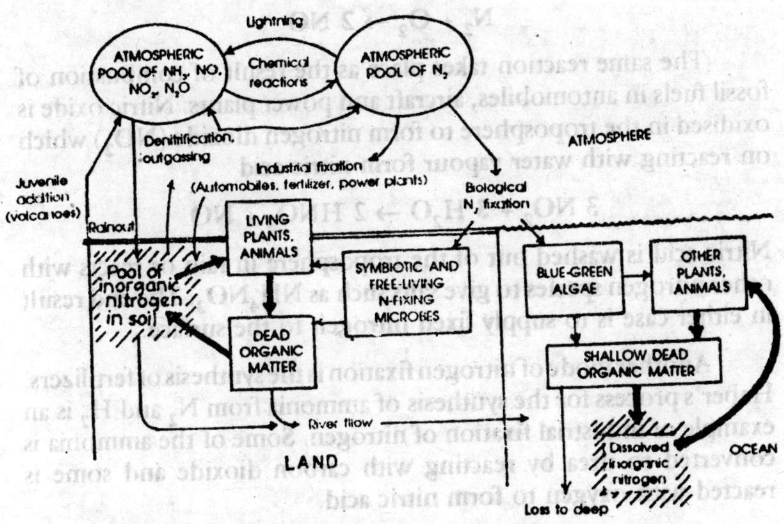

Fig. B.3 : The Global nitrogen cycle. The largest flows are denoted by the dark arrows

transported to the stratosphere where it is both photochemically reduced to N_2 and O, and oxidized by combination with atomic oxygen to nitric oxide (NO). The net reaction is

$$2N_2O \rightarrow N_2 + 2\,NO$$

Nitric oxide is involved in the destruction of ozone (O_3) in the stratosphere through the following reaction :

$$NO + O_3 \rightarrow NO_2 + O_2$$

$$O_3 + h\nu \rightarrow O_2 + O$$

$$NO_2 + O \rightarrow NO + O_2$$

The net reaction is

$$2O_3 \xrightarrow{h\nu} 3\,O_2$$

Thus, there is a link in the biological emissions of N_2O from the soil and the rate of ozone destruction in the stratosphere.

Nitric oxide also enters the troposphere through volcanic emissions. Another source is fixation from atmospheric nitrogen gas by lightening, where the lightning provides the high energy required to drive the reaction

$$N_2 + O_2 \rightarrow 2\ NO$$

The same reaction takes place as the result of combustion of fossil fuels in automobiles, aircraft and power plants. Nitric oxide is oxidised in the troposphere to form nitrogen dioxide (NO_2) which on reacting with water vapour form nitric acid

$$3\ NO_2 + 3\ H_2O \rightarrow 2\ HNO_3 + NO$$

Nitric acid is washed out of the troposphere in rain or reacts with other nitrogen species to give salts such as NH_4NO_3. The end result in either case is to supply fixed nitrogen to the surface.

Another mode of nitrogen fixation is the synthesis of fertilizers. Haber's process for the synthesis of ammonia from N_2 and H_2 is an example of industrial fixation of nitrogen. Some of the ammonia is converted to urea by reacting with carbon dioxide and some is reacted with oxygen to form nitric acid.

The Sulphur Cycle

The sulphur cycle represented in Fig. B.4 has certain important dimensions: the essential role of sulphur in the structure of proteins; the circumstance that the main gaseous compounds of sulphur are toxic to mammals; the fact that sulphur compounds are important determinants of the acidity of rainfall, surface water and soil; and the possibility that sulphur compounds may play a role in influencing the amount of molecular oxygen in the atmosphere in the long term.

The complexity of the sulphur cycle arises mainly from the large number of oxidation states the element can assume. Some of the principal compounds and groups in which sulphur participates are given in Table B.3.

Many transformations among the different oxidation states of sulphur are carried out by bacteria. The main biological transformation are given in Table B.4. Essentially all of the bacterial activity probably takes place in wet media—moist soil, samps, take shores and the open water of lakes and oceans. The presence of dissolved oxygen favours decomposition of organic sulphur to form sulphates; the

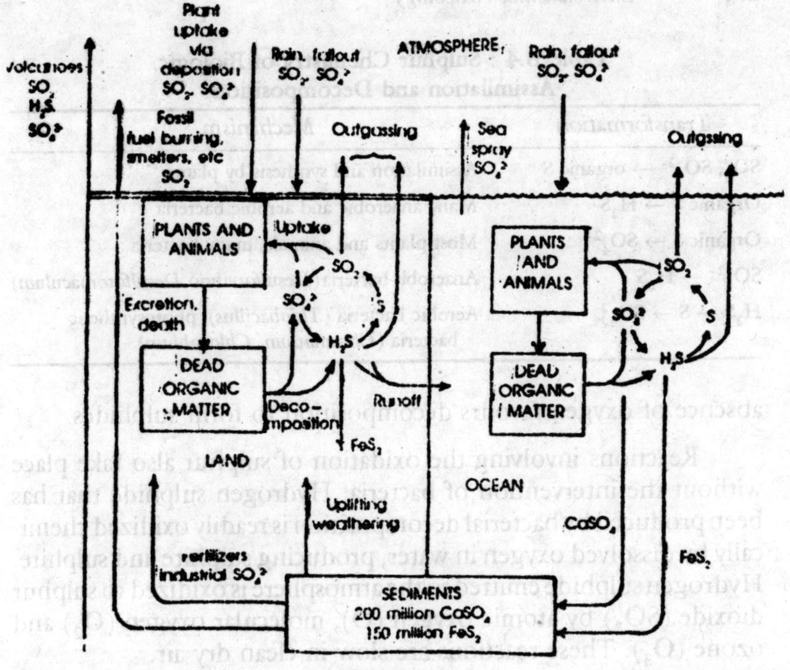

Fig. B.4 : The global sulphur cycle

Table B.3 : Chemical Forms of Sulphur

Formula	*Name*	*Oxidation number*	*Comments*
H_2S	Hydrogen sulphide	–2	"Rotten egg" gas, extremely toxic
HS^-	Hydrosulphide ion	–2	Constituent of amino acids
S^{2-}	Sulphide ion	–2	Forms insoluble compounds with metals
S_2^{2-}	Disulphide ion	–1	Plays crucial role in stiffening protein
S_2,S_6,S_8	Elemental sulphur	0	Crystalline solid
SO_2	Sulphur dioxide	+4	Colourless, toxic gas
H_2SO_3	Sulphurous acid	+4	Weak acid from SO_2 plus water
SO_3	Sulphur trioxide	+6	Gas from oxidizing SO_2 in air
H_2SO_4	Sulphuric acid	+6	Strong acid from SO_3 plus water
SO_4^{2-}	Sulphate ion	+6	Forms many compounds in atmosphere and soil

Table B.4 : Sulphur Chemistry of Biologic Assimilation and Decomposition

Transformation	*Mechanism*
SO_2, SO_4^{2-} → organic S	Assimilation and synthesis by plants
Organic S → H_2S	Many anaerobic and aerobic bacteria
Organic S → SO_4^{2-}	Most plants and animals, many bacteria
SO_4^{2-} → H_2S	Anaerobic bacteria (*Desulforvibrio, Desulfotomaculum*)
H_2S → S → SO_4^{2-}	Aerobic bacteria (*Thiobacillus*), photosynthetic bacteria (*Chromatium, Chlorobium*)

absence of oxygen favours decomposition to form sulphides.

Reactions involving the oxidation of sulphur also take place without the intervention of bacteria. Hydrogen sulphide that has been produced by bacterial decomposition is readily oxidized chemically by dissolved oxygen in water, producing sulphite and sulphite. Hydrogen sulphide emitted to the atmosphere is oxidized to sulphur dioxide (SO_2) by atomic oxygen (O), molecular oxygen (O_2) and ozone (O_3). These reactions are slow in clean dry air.

Sulphur dioxide is further oxidized in the atmosphere to form sulphur trioxide and various sulphates, including sulphuric acid. Photochemical oxidation of SO_2 occurs but is too slow (0.1 percent per hour) to account for the observed lifetime of SO_2 in the atmosphere (minutes to days). A much more important pathway is the dissolving of SO_2 in atmospheric cloud and water droplets to form sulphurous acid (H_2SO_3), and its oxidation to sulphuric acid (H_2SO_4). The oxidation is sped up by various metal salts dissolved in the droplets, which serve as catalysts.

The Phosphorus Cycle

Phosphorus is a manor constituent of biological membranes. Many animals also need large quantities of this element to make shells, bones and teeth Phosphorus probably is the limiting nutrient in more circumstance than any other element because of its scarcity in accessible form in the biosphere.

Two chemical properties of phosphorus are responsible for this natural scarcity. One is that phosphorus does not form any important gaseous compounds under conditions encountered in the environment. The second is the insolubility of the salts formed by the phosphate anion PO_4^{3-} and the common cation Ca^{2+}, Fe^{2+}, and Al^{3+} The lack of gaseous compounds deprives the phosphorus cycle of an atmospheric pathway linking land and sea. Its cycle unlike those of carbon and nitrogen is a **sedimentary cycle**. That phosphate forms insoluble compounds with constituents of most soils retards its uptake by plants and slows its removal and transport by surface water and ground water.

Bulk of the phosphorus present on the earth is integrated in rocks, soil or sediments and distributed fairly uniformly over the entire earth. In the course of the earth's history, deposits have formed at various points of the phosphorus-rich mineral apatite, and these have been used as natural reserves for the steadily rising phosphorus requirements of man.

Energy is required in connection with oil the processes of life in the biosphere, and is converted by green plants from solar radiation into chemical energy. Of fundamental importance in the conversion process is adenosine triphosphate (ATP), a nucleotide consisting of one molecule of the base adenine, one molecule of the sugar ribose, and three molecules of phosphoric acid. As ATP is formed readily by the supply of energy from adenosine monophosphate (AMP) and adenosine diphosphate (ADP) and phosphoric acid, and decomposes again easily upon the release of the energy, it is a suitable compound for transferring energy for metabolic reactions or for storing energy released by decomposition reactions. In its combination with ATP, phosphorus is to be regarded as a key substance in life.

Phosphorus is obtained by plants out of the ground as a phosphate ion. Owing to the poor solubility of phosphorus in water, only a few phosphate ions are available in solution in the ground. The availability of phosphorus is heavily dependent on the soil acidity (pH value) and the form of phosphorus in the soil, in acid soils, iron and acid phosphates dissolve less easily as the decline in the pH value increases, if the ground is basic (alkaline), calcium phosphates are formed, the solubility of which declines as the pH value rises. Ground phosphates are available to plant life in large quantities when the soil reaction is neutral.

If the plants or parts of them die, the organic substance is returned to the soil and decomposed there by phosphatizing bacteria and fungal. In the process, part of the phosphorus incorporated in the plants will be absorbed by the microorganisms, and the rest will be released and pass into the ground. According to the soil acidity and the supply of aluminium and calcium ions, the phosphorus will again be established in the form of iron, aluminium or calcium phosphate, if the plants are consumed by animals, phosphorus contained in them will be ingested by the animal organism and ultimately be returned to the soil in excrement or with the carcass. Thus, phosphate cycles round and round on land from plants to animals and back again. Land ecosystems preserve phosphorus efficiently, since both organic and inorganic soil particles absorb phosphate, providing a local reservoir of this element. In an undisturbed ecosystem, the intake and loss of phosphorus are small compared with the amounts of phosphorous that are internally recycled in the day to-day exchange among plants and animals. Some phosphorus is inevitably lost by leaching and erosion of the soil into streams and rivers. When an ecosystem is disturbed, for instance by mining or farming, erosion can become so significant that large quantities of phosphorus and other minerals are washed away. When phosphorus reaches the ocean, it encounters other minerals in the water and reacts with some of them. Since most phosphates are not very soluble in water, they form insoluble compounds and eventually settle to the ocean floor. One reason for the infertility of open oceans is their low phosphorus content.

In the atmosphere, phosphorus is largely combined with aerosols. Its displacement in this form is of consequence only in dust- and sand-storms and volcanic eruptions. When the air turbulence subsides, the dust particles settle down again, Phosphorus normally gets into stream and river currents only through erosion of soil particles containing phosphorus.

The links between the terrestrial and oceanic parts of the phosphorus cycle are very weak. Little dissolved phosphorus is carried by rivers, because of the Ice solubility of phosphorus soils (the amount carried in suspended soil particles by erosion is about ten times as large). Aside from the slow link provided by the sedimentary cycle, the only sea-to-land transport is in fish and shellfish harvested from the sea and consumed on land, and in the excrement that fish eating-sea birds deposit on land. The only other way that phosphorus can return to land is by way of extremely slow geological processes in which sea floor sediments rise form the sea to form islands or continents. Soil erosion robs an ecosystems of its phosphorus, and it may take thousands or tens of thousands of years to recoup this loss through the weathering rocks, Phosphate is much in demand as a fertilizer, but the richest and most easily mined deposits of phosphate are being depleted rapidly.

Phosphorus enters the water by several anthropogenic pathways. It is applied in fields as fertilizer and is present in sewage and in the waters from livestock feeding. Phosphates added to detergents to improve their "Cleaning power" become the ingredient of municipal waste water and are not removed effectively from sewage by primary or secondary treatment plants.

Phosphorus is the limiting nutrient in the growth of aquatic plants due to the less-soluble nature of phosphorus compounds and thus it does not move readily through soil into the waterways so its concentration is less in most natural bodies of water.

Because phosphorus compounds are so tightly bound to the soil, even very heavy fertilization does not lead to the leaching of phosphorus into ground water and surface water. Fertilizer phos-

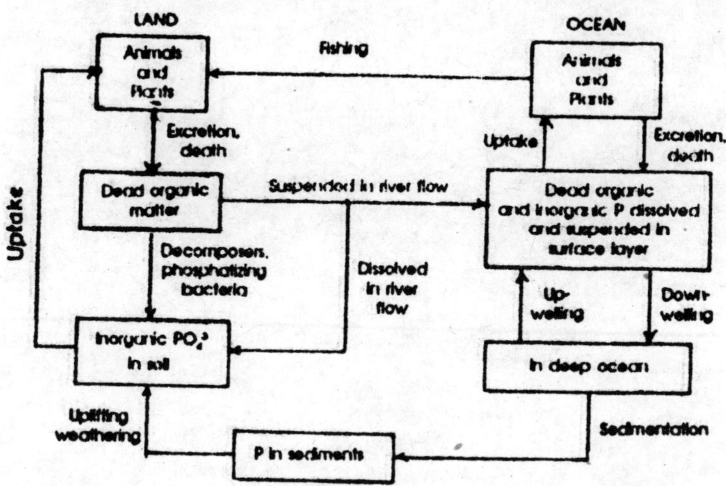

Fig. B.5 : The phosphorus Cycle

phorus that reaches waterways does so almost entirely by being carried along in eroded soil particles. Even in the water, most of that phosphorus remains in suspension (not in solution) so it is not readily accessible for use by aquatic plants. By far the largest source of soluble phosphorus in waterways is municipal sewage, which contains phosphorus both from excrement and from detergents. Many examples of cultural entrophiation of lakes, due mainly to phosphorus addition in sewage have been reported.

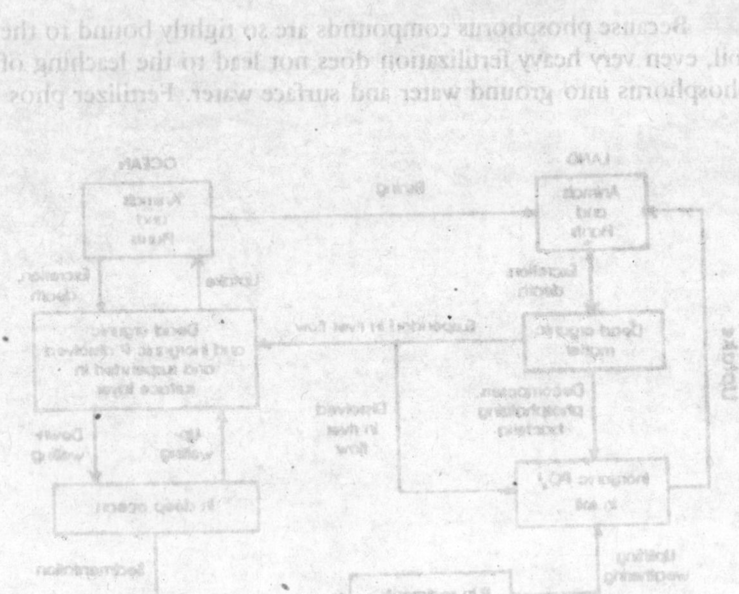